Human Generativity: An Introduction to the Human Sciences

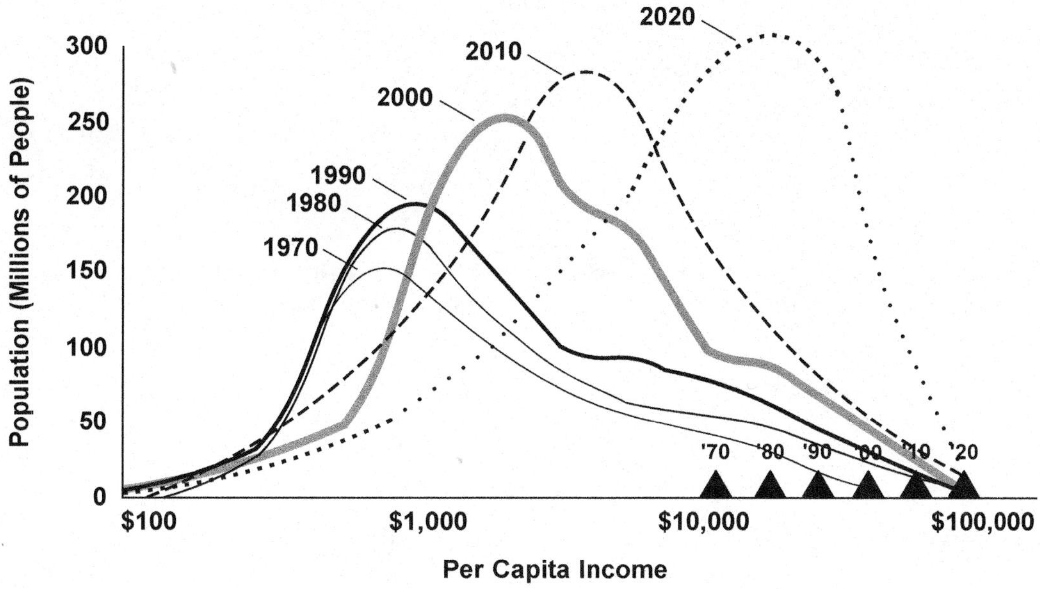

▲ = U.S. Leadership

Robert R. Carkhuff, Ph.D.

HRD Press, Inc. ❖ Amherst ❖ Massachusetts

Published by: HRD Press, Inc.
22 Amherst Road
Amherst, MA 01002
800-822-2801 (U.S. and Canada)
413-253-3488
413-253-3490 (fax)

ISBN 978-1-61014-292-2

Editorial services by Robert W. Carkhuff
Production services by Jean S. Miller
Cover design by Eileen Klockars
Promotion by Swift Global Media

Human Generativity: An Introduction to the Human Sciences

Robert R. Carkhuff, Ph.D.
Chairman, The McLean Project

Executive Directors

• **Physical Sciences**	Tom Kakovitch, D.Sci.
• **Human Sciences**	John R. Cannon, Ph.D. Sharon Fisher, MSW
• **Individual Generativity**	Don Benoit, M.Ed. Richard M. Pierce, Ph.D.
• **Informational Generativity**	George Paley, D.Sci.
• **Organizational Generativity**	Christopher J. Carkhuff, D.Sci.
• **Community Generativity**	George Banks, Ph.D. Karen Banks, Ph.D.
• **Cultural Generativity**	Al Cook, Ph.D. Rob Owen, MBA
• **Economic Generativity**	Hernan Oyarzabal, D.Sci.
• **Parametric Measurement**	Richard C. Sprinthall, Ph.D.
• **Software Development**	Peter Armstrong, B.S. Kenneth Pratt, B.S.

Human Generativity:
An Introduction to the Human Sciences

Contents

About the Author

Among the most-cited scientists of the 20th century and already the most prolific in the 21st century, Robert R. Carkhuff is the author of *The Human Sciences.* His full biography and body of work may be viewed on his websites:

www.carkhuff.com

www.mcleanproject.com

Carkhuff boldly confronts our current socioeconomic crises:

"Generativity is the solution!"

"What is the question?"

Preface

Generativity and Equitability

Take another look at the graphic image on our cover! It is the most exciting and inclusive representation of our times. It depicts both the successes and the failures of our efforts following World War II.

First, the successes! Our long-term prospects are found in the demonstration of our **Global Domestic Product.** Data-based, our image projects the gigantic **"Tsunami Wave of Economic Growth"** in which the world has participated for over half a century. Led by **"The New Capitalism"** generated by America's free enterprise model, this wave has generated prosperity in which all peoples from all cultures and nations have shared.

Second, the failures! The critical problems are found in the awesome disparity or inequity with which our prosperity is distributed. While all parties have shared in the benefits of the New Capitalism, people toward the tail of the wave have received disproportionately far less than those at the front have received. The disparity is compounded by the hard-wired conditioning of the disenfranchised peoples themselves to pursue conventional ways of behaving.

This book is about **generativity—the capacity to create whole new ways of doing things in the spirit of America's initiative definition of economic leadership.** Generativity is the only answer to our problems of both prosperity and poverty!

Inevitably, this is a war fought for the future of humankind. The opponents are defined in sharp relief:

- Socioeconomic growth through generativity
- Socialistic stasis through planning

Ultimately, this is about the dedicated use of human brainpower:

- To the tsunami wave of economic growth
- Or the tsutsugamushi mite hollowing out our brainpower

We may view the linear representation of the cause and effects of growth and prosperity in the graphic on the opposite page:

- America leads the way with its traditions of cultural relating, participative governance, and entrepreneurial free enterprise.
- The other nations follow (and in some cases exceed the American paradigm) and everyone benefits.

This American model has now been put at risk by Communistic and Socialistic economics (*The Black Book of Communism,* Harvard University Press, 1999). Having failed everywhere else, socialistic thinking seeks to bring the American model to its knees by making "entitlement" available to the majority rather than entrepreneurism to the minority.

To be sure, while the American model has not gone far enough, it should be built upon by the next two great influences upon civilization:

- The New Capitalism
- Generativity

Human history—if we are to continue one—will record our decisions.

God will judge us!

RRC

January 2013
McLean, Virginia

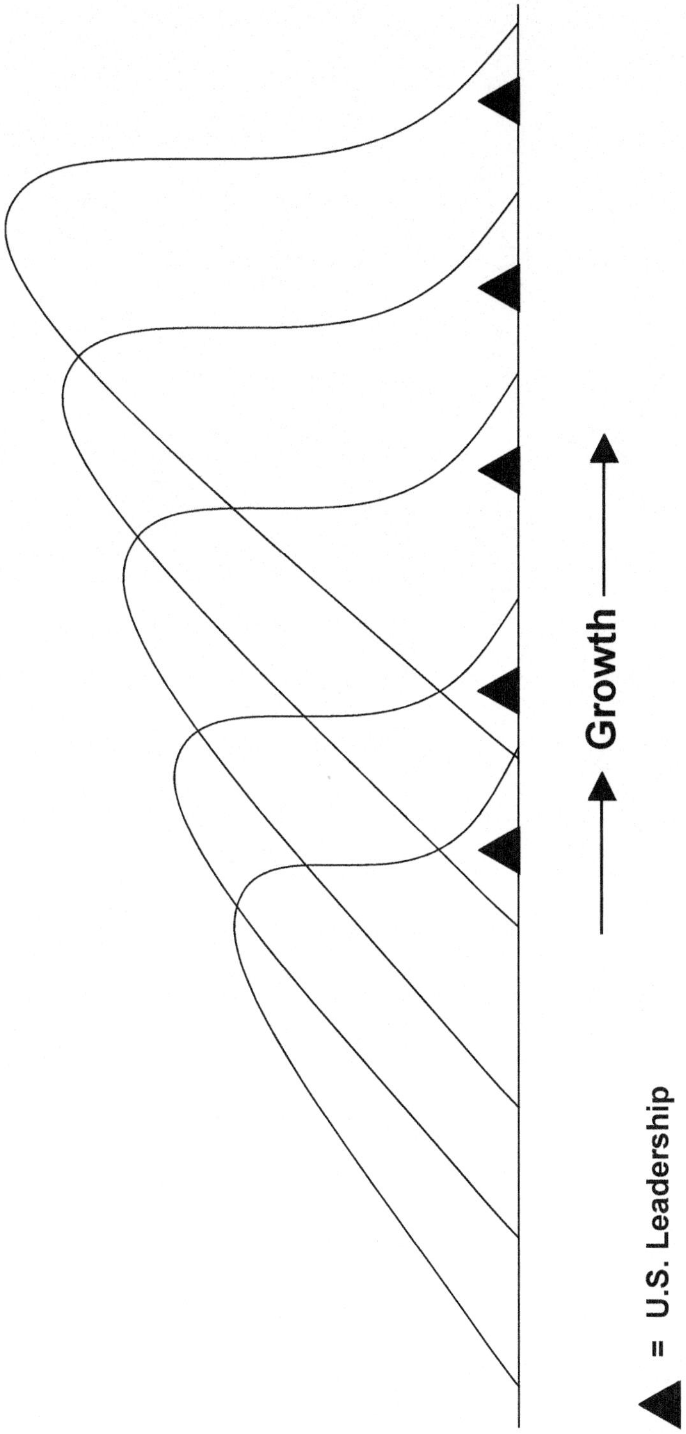
Growth
▲ = U.S. Leadership

Foreword

Eighteen months after being introduced to **"The Age of the New Capitalism"** in a personal audience with Carkhuff, Pope John Paul II issued the **Papal Encyclical, "The New Capitalism."**

The New Capitalism

Can it perhaps be said that, after the failure of Communism, capitalism is the victorious social system, and that capitalism should be the goal of countries now making efforts to rebuild their economy and society?

If by "capitalism" is meant an economic system that recognizes the fundamental and positive role of business, the market, private property, and the resulting responsibility for the measure of production, as well as free human creativity in the economic sector, then the answer is certainly in the affirmative, even though it would perhaps be more appropriate to speak of a "business economy," "market economy," or simply "free economy." (John Paul II, Centesimus Annus, 42.1–42.2, in Miller, *Encyclicals.*)

Source: Carkhuff, R. R. *The Freedom Wars.* HRD Press, 2004

Dedication

To My Generative Mentors

David N. Aspy, Ed.D.
Director Emeritus, National Consortium for Humanizing Education

Bernard G. Berenson, Ph.D.
Professor Emeritus, American International College

B. Richard Bugelski, Ph.D.
Professor Emeritus, SUNY, Buffalo

James Drasgow, Ph.D.
Professor Emeritus, SUNY, Buffalo

John T. Kelly, D.Sci.,
Director Emeritus, A.S.D., IBM, Inc.

Richard C. Sprinthall, Ph.D.
Professor Emeritus, American International College

"Be ashamed to die until you have achieved a victory for humanity."

— Horace Mann

Part I

The Generativity Explosion

1

The Paradigm Shift to Generativity

There is a story told about an old man doffing his hat to a passing funeral procession. As he prayed silently for the victim, his companion commented on how touching his behavior was. The old man replied:

"Yeah, well I was married to her for 50 years, but I made a mistake."

Amidst the tumult of crashing markets and unsustainable bailouts, a similar thought has been expressed by the promulgators of free market enterprise:

"We were wrong!"

From the surrounding gurgles of Greenspan and Mandelbrot to the condemning silence of the intellectual elite, they have all expressed themselves in unison:

"We thought we were right until we found we were wrong!"

So what happened? Where did they go wrong?

The answer is simple: We chose to believe naively and greedily in the grandiose proposition of the marketplace—that markets are self-adjusting and self-regulating.

The American Tsunami

We are approaching the end of the greatest half-century of socioeconomic growth in the history of humankind. Indeed, it is growth greater than was accomplished in the history of recorded civilization. To be sure, the last five decades have generated more wealth than the 14 million years of human evolution. And America has led this growth with its free enterprise principles and its free trade policy!

Why is it then that we Americans are not celebrating this most awesome accomplishment in human history? Could it be because we find ourselves attenuating in our leadership characteristics even as we are serving to culminate world growth? Could it be that even as we are powering forward and empowering others to do likewise, we are being hollowed out by enemies from within and without?

We may view illustrations of the influence of **The Freedom Functions** upon **World Income Distribution** (see Figure 1). As may be noted, there is significant growth in per capita income for each decade. Not only is growth in income relative for countries but absolute for the world.

These world economic curves are made up of individual country curves. As may be noted, the curves build to a crescendo. Like a gathering tsunami, they follow the leader. Like a gathering tsunami, they project to build forever.

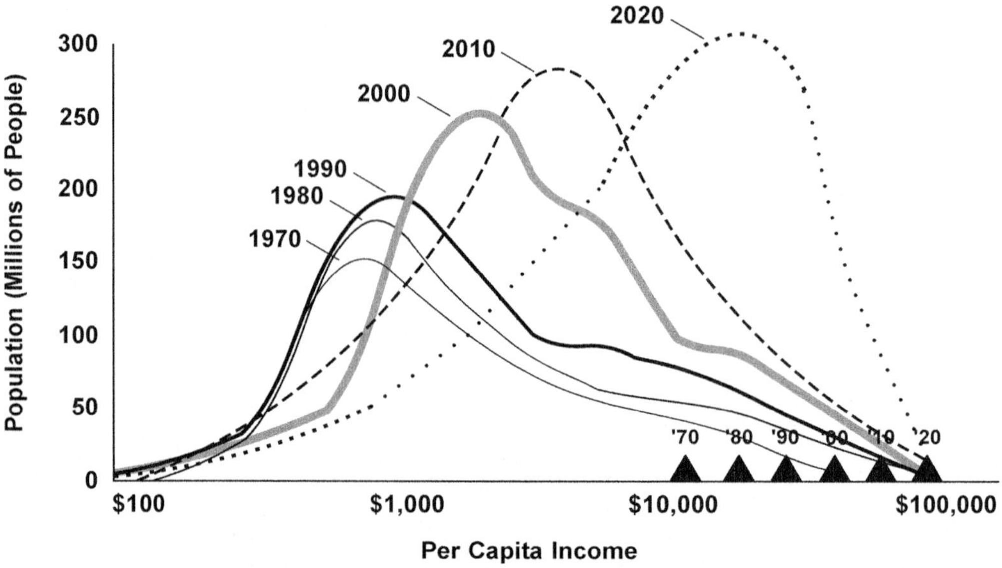

Figure 1. World Per Capita Growth Projected by Decades

A tsunami is a huge tidal wave caused by seaquakes and volcanic eruptions. Conceived as an economic wave, it promises huge benefits. Why are we not celebrating the projected benefits? Could it be because we fear the projected costs to our freedoms, our way of life, our way of learning, our way of working, our labor?

Could it be that we Americans fear the tsutsugamushi of China—microorganisms transmitted by mites that cause the infectious disease of rickets? Could it be that we Americans fear being hollowed out as we play the parasitical host endorsed by our own leaders, including especially our president?! Could he be wrong? Let us explain.

The Generativity Paradigm

We are the people who have been able to postdict our origins. We are also the people who can predict our futures. Relating to our experiences, empowered by our generative processing systems, we are free to define our own empowerment programs and to generate our own futures intentionally.

The generativity paradigm may be viewed in Figure 2. As may be noted, it characterizes all elements of human endeavor with the "generativity" adjective: individuals, organizations, communities, cultures, economics. In other words, the driving force is *Generativity Processing.*

Simply stated, "generativity is the human brainpower necessary to create new and more productive ideas." It means "seeing beyond the high beams" and, indeed, "beyond the moon beams."

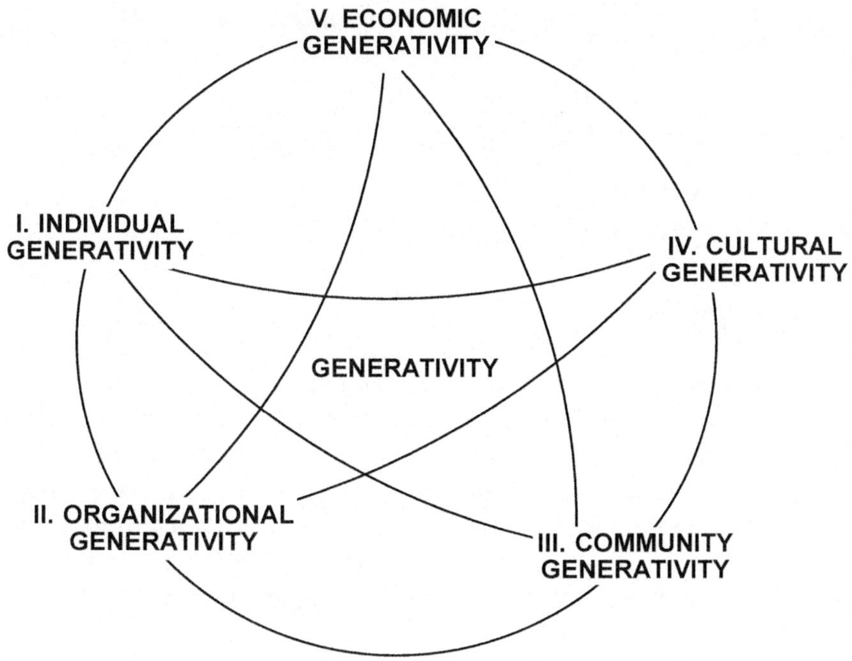

Figure 2. The Generativity Paradigm

After researching the dimensions of generativity, we are now able to empower individuals and organizations to systematically apply the generativity ingredients. These individual and organizational generativity systems are nested or housed in community generativity, which provides the new capital foundation for all human endeavors. In turn, the community generativity is nested in cultural generativity, which actualizes prosperity, participation, and peace in our time. Finally, all the generativity systems are dedicated to economic generativity, which emphasizes the interdependent relationships between generators, innovators, and commercializers within the marketplace. These generativity systems are the sources of *economic capital development.*

The issue before us today is not how to invest another trillion dollars in worker employment or consumer confidence. The issue is how do we achieve the highest levels of generativity leading to the highest levels of civilization? The problem is getting to this civilized place.

The generativity paradigm may be viewed as follows:

- Individual generativity developing human capital and information capital
- Organizational generativity developing organization capital
- Community generativity developing community capital
- Cultural generativity developing cultural capital
- Economic generativity developing economic capital

Developmentally, each stage of generativity is "nested" or housed in the higher-order stages:

- Networks of individuals that comprise organizations
- Networks of organizations that comprise communities
- Networks of communities that comprise cultures
- Networks of cultures that comprise economies

This means that economic generativity comprises cumulative networks of cultures that process generatively.

What makes this generativity paradigm different from all other sociopolitical proposals is that it is derived from actionable architecture and operational technologies. In other words, this platform is based on projects that have been validated and are, therefore, predictable in their effects.

Generativity is the only stimulus package that will work for America! Generativity in all areas of human endeavor, alone, can bring America back to its leadership role in the prosperous world economy. Generativity must be the foundation for all platforms that define all leadership priorities in America now, at the beginning of the 21st century.

2

The Generativity Systems

The miracle of generativity may be seen in sharpest relief in relation to our freedom. All of us encounter moments in our lives when we must think "outside of the box." We may have lost our jobs and our incomes, or our opportunities in other areas such as studying for advancement. The whole idea of finding ourselves "boxed in" implies a loss of degrees of freedom.

In our lifetimes, authority usually devolves from the top down. Often, we are offered the "go–no go" solutions: either do the work or leave! Or, as our current economic crisis dictates: either be regulated or deregulated! This gives us the boxed-in feeling.

Generativity offers another way. Indeed, it offers many other ways—potentially an infinite number of ways—but only if we care enough to explore them!

Generativity is a "thinking process." Functionally, it creates new and better solutions to all predicaments by expanding and narrowing alternative courses of action. We may participate in generative processing—individually or interpersonally with others, on a team in an organization, in an organization in a community, in a component of a community in a culture, in an aspect of the economy.

In the Generativity Systems, I am going to introduce you to the processes of generativity that empower us as individuals and as units of people to create new and more productive solutions to our predicaments.

The miracle of generativity is that it offers solutions from both the bottom up as well as the top down. Indeed, these solutions should occur simultaneously.

In making our presentations, we are going to begin developmentally with individual generativity.

Individual Generativity

The first and most basic process of generativity is *individual generativity* (see Figure 3). Individual generativity empowers us as individuals to generate solutions to all crises, and individual generativity empowers us to engage in generative processing as follows:

- **R^1 – Relating** to the images of the phenomenon
- **R^2 – Representing** to improve images of the phenomenon
- **R^3 – Reasoning** to generate the most productive images of the phenomenon

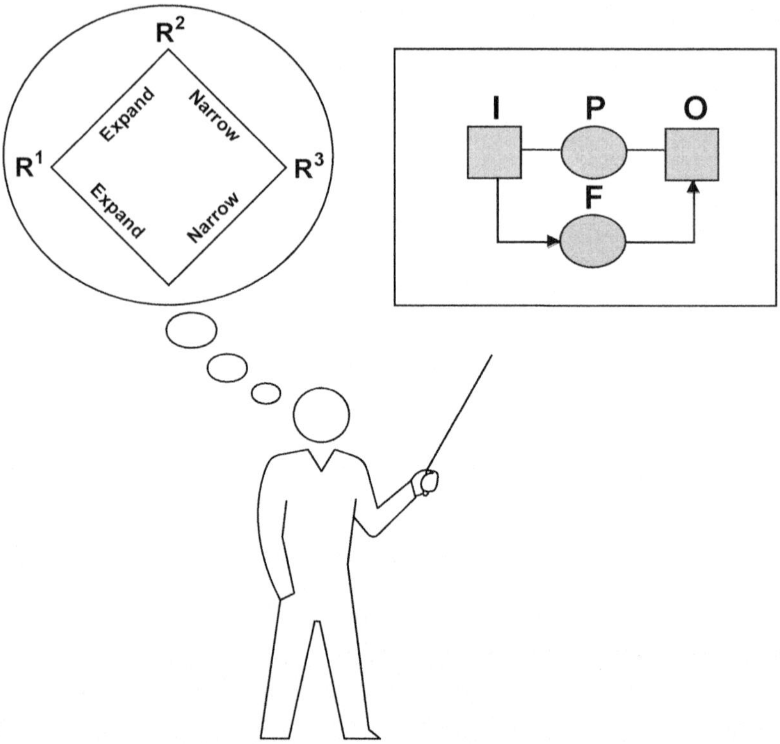

Figure 3. Individual Generativity

As may be viewed in the generativity process, we expand alternatives before narrowing images. We may go on to share our images with others, after which we may "merge" images that are most preferred for meeting our requirements and capacities. In the illustration, we see the individual who, after expanding and narrowing images, generates the preferred image: a system with inputs, process, outputs, and feedback (I–P–O–F).

The net effect of individual generativity is to develop **Human Capital Development (HCD)** and **Information Capital Development (ICD).** In essence, individual generativity is the synergistic or mutual growth-producing relationship between Human and Information Capital Development:

Individual Generativity → (HCD ↔ ICD)

Organizational Generativity

Since I began my career, I have often been working with individuals who were victims left in the wake of independent and impersonal organizations. I was initially suspicious of organizational processing. Then I discovered in my research that organizations are merely networks of individuals that, just as with individuals in teams or units, may process "for better or worse."

The second process of generativity, then, is to relate individual generativity to *organizational generativity* (see Figure 4). Within the organization, we relate the individual processors "interdependently" with each other. Thus, at both individual work stations and within units, the individuals are processing independently and interdependently. In this manner, the organization is able to accomplish all of the functions to accomplish its mission.

The net effect of organizational generativity is **New Capital Development** or **NCD:**

- **MCD**—Marketplace Positioning
- **OCD**—Organizational Alignment
- **HCD**—Human Processing
- **ICD**—Information Modeling
- **mCD**—Mechanical Tooling

In essence, then, organizational generativity is the source of New Capital Development (NCD):

Organizational Generativity → NCD

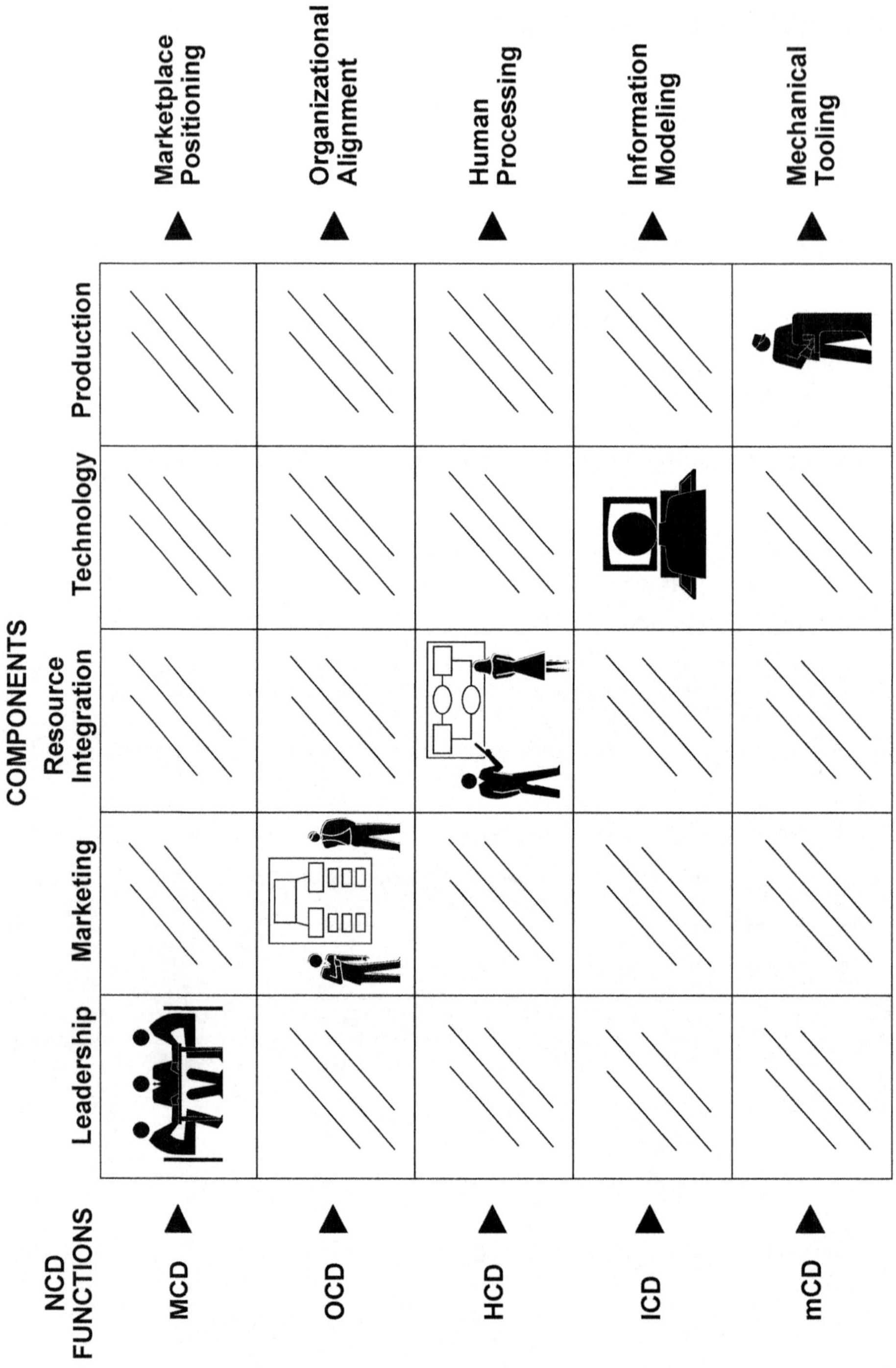

Figure 4. Organizational Generativity

Community Generativity

I personally have spent much of my professional life in community resource development. Just as with organizations, I have found that communities are simply networks of organizations: homes, schools, colleges, governments, businesses. Relatedly, just as with organizations, I have found that community components can become interdependent generative processors.

The third process of generativity, then, is to relate generative organizations in *the generative community* (see Figure 5). Within the community, we relate the organizations interdependently with each other. Thus, all components process interdependently within, between, and among organizations. In this manner, the community is able to accomplish all of the functions necessary to accomplish its mission of **Community Capital Development** or **CCD.**

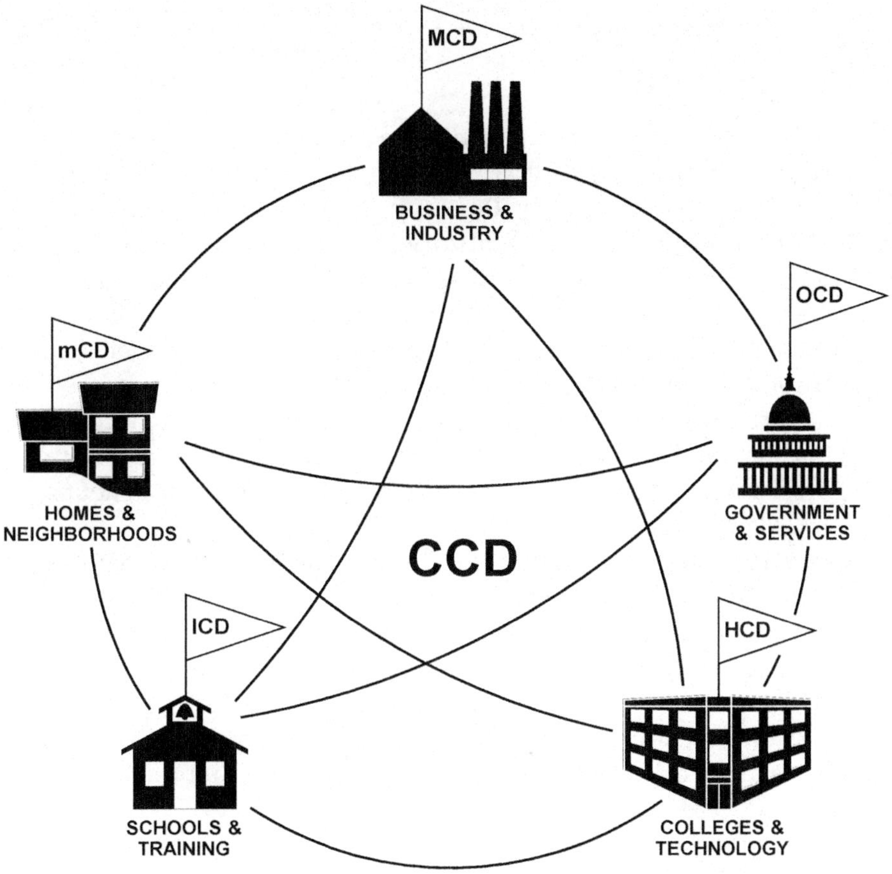

Figure 5. Community Generativity

The net effect of the generative community is **Community Capital Development** or **CCD.** Community components specialize in New Capital Development:

- Homes emphasize Mechanical Capital Development or **mCD.**
- Schools emphasize Information Capital Development or **ICD.**
- Colleges emphasize Human Capital Development or **HCD.**
- Governance emphasizes Organizational Capital Development or **OCD.**
- Business emphasizes Marketplace Capital Development or **MCD.**

In essence, then, the generative community is the source of Community Capital Development:

Community Generativity → CCD

Cultural Generativity

The secrets of *cultural generativity* came to me in my maturity. The fact is that we in America have lived in the most generative culture in the history of civilization, and we have derived the benefits directly. Our 50 states have lived in peace unlike any other similarly positioned states in the history of the world. The foundations of our democratic governance have delivered participation on levels unknown to any other citizens in the world. Our entrepreneurial capitalism has accounted for by far the greatest technological advances and job creation in the last five decades, dwarfing the prosperity of the entire history of civilization.

In our research, we have discovered that cultures are networks of communities committed to common functions. Those cultures dedicated to cultural relating, participative governance, and free enterprise thrive on all indices of benefits: peace, participation, and prosperity.

The fourth process of generativity, then, is to relate generative communities to cultural generativity (see Figure 6). Within the culture, all communities process interdependently to accomplish the cultural mission. For example, the American culture has been characterized by collaborative relating, democratic governance, and entrepreneurial enterprise.

The net effect of cultural generativity is **Cultural Capital Development** or **CCD′.** In essence, then, cultural generativity is the source of Cultural Capital Development or CCD′:

Generativity Culture → CCD′

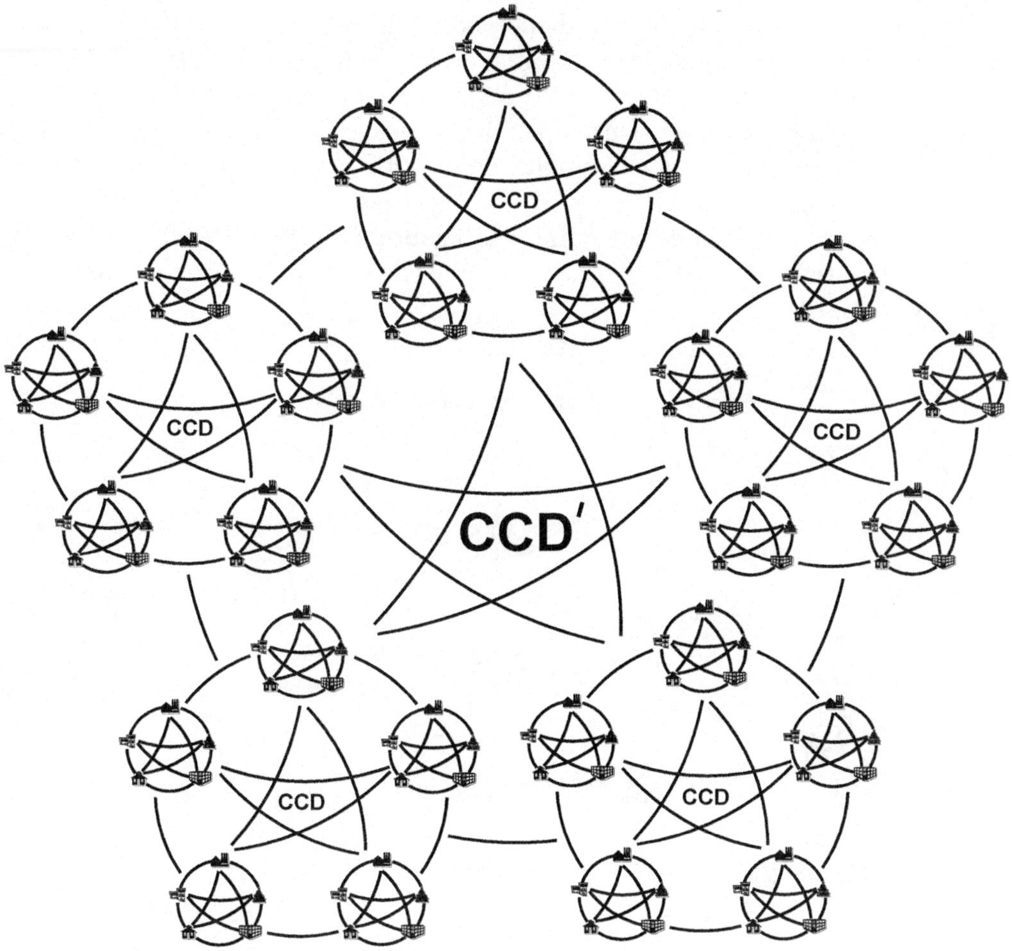

Figure 6. Cultural Generativity

Economic Generativity

The process of *economic generativity* came to me in my work with private sector corporations and public sector communities. Those entities that practiced generative processing in every area and at every level of the organization and communities became generative cultures. An economy is simply a network of cultures dedicated to a common socioeconomic mission.

As we will see, the secret to any economy's success is positioning for comparative advantage in the marketplace. This means establishing that we have products and services that our trading partners value because of their quality, cost, and timeliness.

The fifth process of generativity, then, is to relate cultural generativity to economic generativity (see Figure 7). Within the economy, the cultures relate interdependently with each other. Thus, all cultures process interdependently to accomplish their economic mission. For example, the American economy has historically supported entrepreneurial capitalism with democratic governance and collaborative relating.

The net effect of economic generativity is **Economic Capital Development** or **ECD.** In essence, then, economic generativity is the source of **Global Socioeconomic Development:**

Economic Generativity → ECD

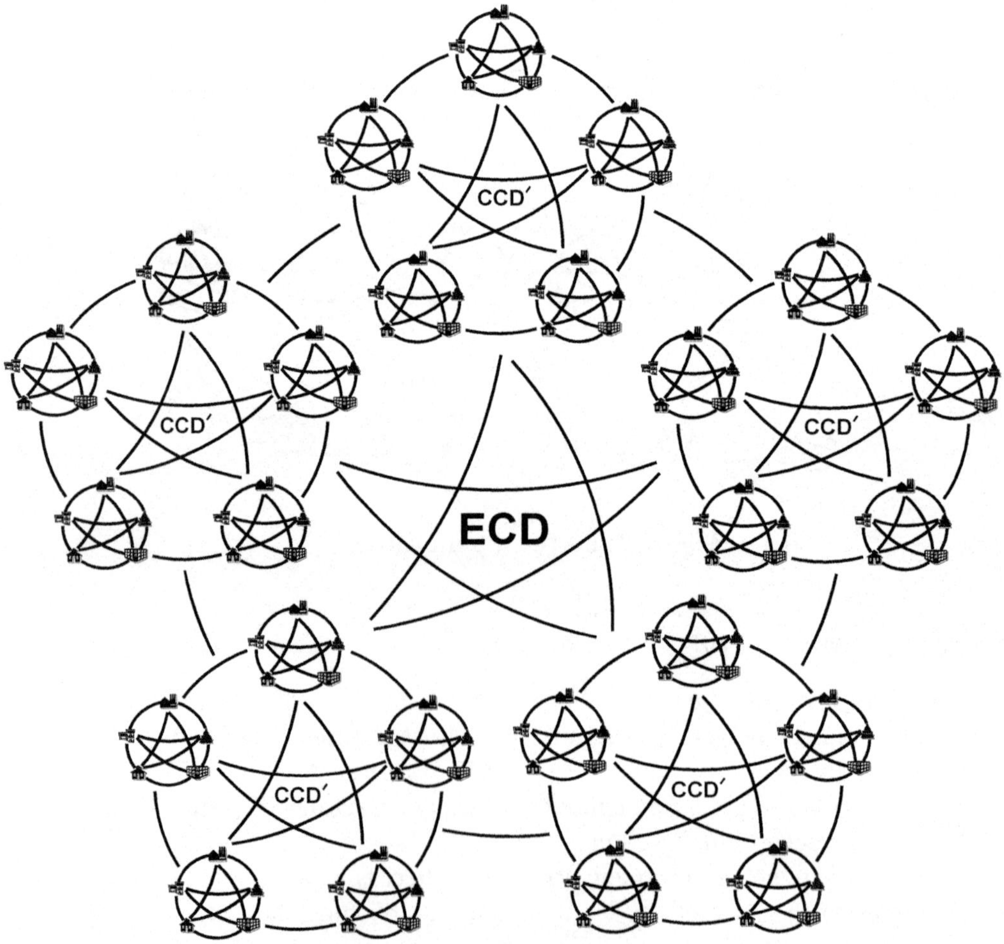

Figure 7. Economic Generativity

Part II

The Sources of Generativity

3

The Freedom Functions

Freedom is the exercise of choice in the face of definable conditions. We have chosen to define freedom for individuals in terms of the repertoire of processing responses. The same definitions hold under differing conditions for organizations, communities, cultures, markets, and, yes, nations. All define freedom by their ability to process the conditions of their existence, often in the moment, always in the crisis. Freedom may be defined operationally by the **Freedom Matrix** (see Figure 8). This is a two-dimensional matrix comprising interacting scales for values and requirements.

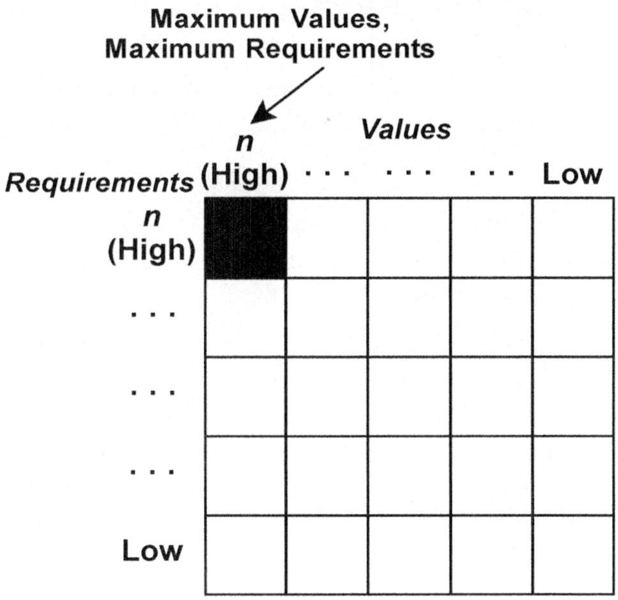

Figure 8. The Freedom Matrix

To better understand the Freedom Matrix, individuals may think of themselves going for a job interview:

- **Values** are the things that people would like to acquire by attaining the job—things like pay, location, comfort, security, and fulfillment (Values 1 through *n*).

- **Requirements** are the things that the job wants from the people they hire—things like showing up, having skills, performing tasks, being productive and maybe even profitable (Requirements 1 through n).

In effect, the job's values become the individual's requirements. Similarly, the individual's values become the job's requirements: recruiters position the jobs to be attractive to people who are attractive to them.

It is similar for all other levels of conditions. **All entities relate values and requirements in healthy attempts to maximize meeting both at the highest levels.**

Nations seeking to join the global village or enter the global marketplace have similar choices. They have values for joining: participation, peace, prosperity. In turn, the market presents requirements for being chosen: participation, peace, prosperity! In this instance, the values and requirements focus upon the same dimensions. The differences may be found in the standards of the freedom functions.

In this context, then, we define *freedom* as a function of the following:

> **The extraordinary quantity and quality of processing systems of all kinds, dedicated to elevating the rational principles of a continuously changing civilization.**

In turn, we define *totalitarianism* as a function of the following:

> **The surrender of processing potential to the will of the necessarily limited "master" who promises the illusion of stability and the delusion of competence.**

Again, freedom is processing: a function of our repertoire of generative responses to changing stimuli. Totalitarianism is conditioning: a function of our conditioned responses to unchanging stimuli.

The freedom functions define the **freedom architecture** for the standards of freedom for all nations. They emphasize the levels of freedom that nations must achieve in order to meet the requirements for participation, contribution, and leadership in the global village and its marketplace (see Table 1).

Table 1. The Freedom Architecture

	Stages of Freedom-Building	Cultural Relating	Participative Governance	Economic Enterprise
GROWTH ↑	Leader	Free Interdependent	Free Democratic	Free Enterprise
	Contributor	Collaborative	Representative	Capitalistic
	Participant	Independent	Mixed	Mixed
	Observer	Competitive	Authoritarian	Command
SURVIVAL	Detractor	Dependent	Totalitarian	Control

As may be noted, the freedom functions define the dimensions of freedom for nations:

- **Cultural relating** or the degrees of relatedness within, between, and among cultures and nations
- **Participative governance** or the degrees of participation in governance within, between, and among cultures and nations
- **Free enterprise** or the degrees of freedom in economic enterprise within, between, and among cultures and nations

Together, the freedom functions define the freedom architecture for entry and elevation in the 21st century global marketplace.

Cultural Relating

With constructive relating, all human endeavors are probable with a 95% degree of certitude. With constructive cultural relating at high levels, all national initiatives are possible with a similarly high degree of certainty. The plain fact of the matter is that nations cannot trade if they do not relate. Emphatically, nations cannot trade in the global marketplace without relating in the cultural marketplace!

In this context, the levels of cultural relating are defined in the Freedom Model:

Level 5: **Free and interdependent relating** is defined by "mutual processing for mutual benefit" with other cultures and/or nations.

Level 4: **Collaborative relating** is defined by shared goals and common means to achieve the goals with other cultures and/or nations.

Level 3: **Independent relating** is defined by reserving the right and power to compete as well as cooperate with other cultures and/or nations.

Level 2: **Competitive relating** is defined by achieving and maintaining the *competitive edge* with other cultures and nations.

Level 1: **Dependent relating** is defined by dependency in relating with other cultures and nations, including adapting and reacting.

Participative Governance

In the enlightened, modern world, governance is defined by participation. Government without participation restricts decision making to a council or a person of increasingly isolated, totalitarian mind-sets: they make decisions about citizens **for** citizens. And with the acceleration of change, they "know" increasingly less and less, until they "know" nothing about anything! Even when they are overthrown or replaced, one limited set of variability is replaced by another limited set of variability. The only real discrimination in governance is the degree of participation.

In this context, the levels of participative governance are defined in the Freedom Model:

Level 5: **Free and interdependent democratic governance** defined by direct and **mutually beneficial** relations between citizens and decision makers

Level 4: **Representative democratic governance** defined by citizen relations with representatives who represent them with decision makers

Level 3: **Mixed governance** as defined by occasional or intermittent democratic and authoritarian activities

Level 2: **Authoritarian governance** as defined by intentional governmental edicts

Level 1: **Totalitarian governance** as defined by total control of political practices

Economic Enterprise

The key to the highest levels of economic enterprise is the modifier, "free." It is not the trade that defines it. It is the free thinking that generates initiatives to meet consumer needs that defines it. Specialty products make trade attractive. Trade agreements without attractive products are like rental properties without occupants. The plain fact of the matter is that nations cannot trade if they do not have something of value. And they will not have something of value if they cannot generate substance!

In this context, the levels of economic enterprise are defined in the Freedom Model:

Level 5: **Free and interdependent entrepreneurial enterprise** defined by "mutual trading for mutual benefit" with other nations

Level 4: **Capitalistic economics** defined by maximizing financial capital-based investments and minimizing governmental burdens

Level 3: **Mixed economics** defined by occasional or intermittently capitalistic and command practices

Level 2: **Command economics** defined by intentional governmental command practices

Level 1: **Control economics** defined by total governmental control of economic practices

4

The Growth Outcomes

It is widely believed that any indices of freedom are related to everything that is good in the world: participation, intelligence, creativity, performance, and, yes, peace. However, the evidence from research is limited. Moreover, the research does not specify the programs and objectives for elevating freedom in operational terms.

An exception to these beliefs has been the work of the Heritage Foundation that has been the forerunner in assessing the effects of economic freedom. Defining economic freedom as "the absence of government coercion" upon the private sector, the researchers accounted for much of the variance in prosperity through the following measures: trade policy, fiscal burdens, government intervention, monetary policy, capital flows, banking and finance, wages and prices, property rights, regulation, and black market.

In our own work, we have studied the relationship between prosperity and the freedom functions: cultural relating, participative government, economic enterprise. In Figure 9, we present the relationship of our freedom scores and prosperity measured in per capita gross domestic product or GDP.

As may be noted, in the "power curve" there is a high relationship between freedom and per capita GDP: high scores on the freedom functions earn high GDP and vice versa. This means that, as the freedom index moves toward "free," the per capita GDP moves upward many thousands of dollars. Conversely, as the nations move toward "detractor," the per capita indices move toward low earnings, ultimately numbered in hundreds of dollars per year.

The correlation between freedom scores and prosperity indices is high.

- **The correlation between prosperity indices and freedom scores is + 0.75.**
- **This means that more than 50% of the variance in prosperity is accounted for by the freedom functions: cultural relating, participative governance, free enterprise.**

The correlation is not a perfect one-to-one correspondence: one movement up in earnings for one movement over in freedom. That is because of other factors; for example, some nations that are totalitarian may also be rich in resources such as energy—so they get "middling earnings."

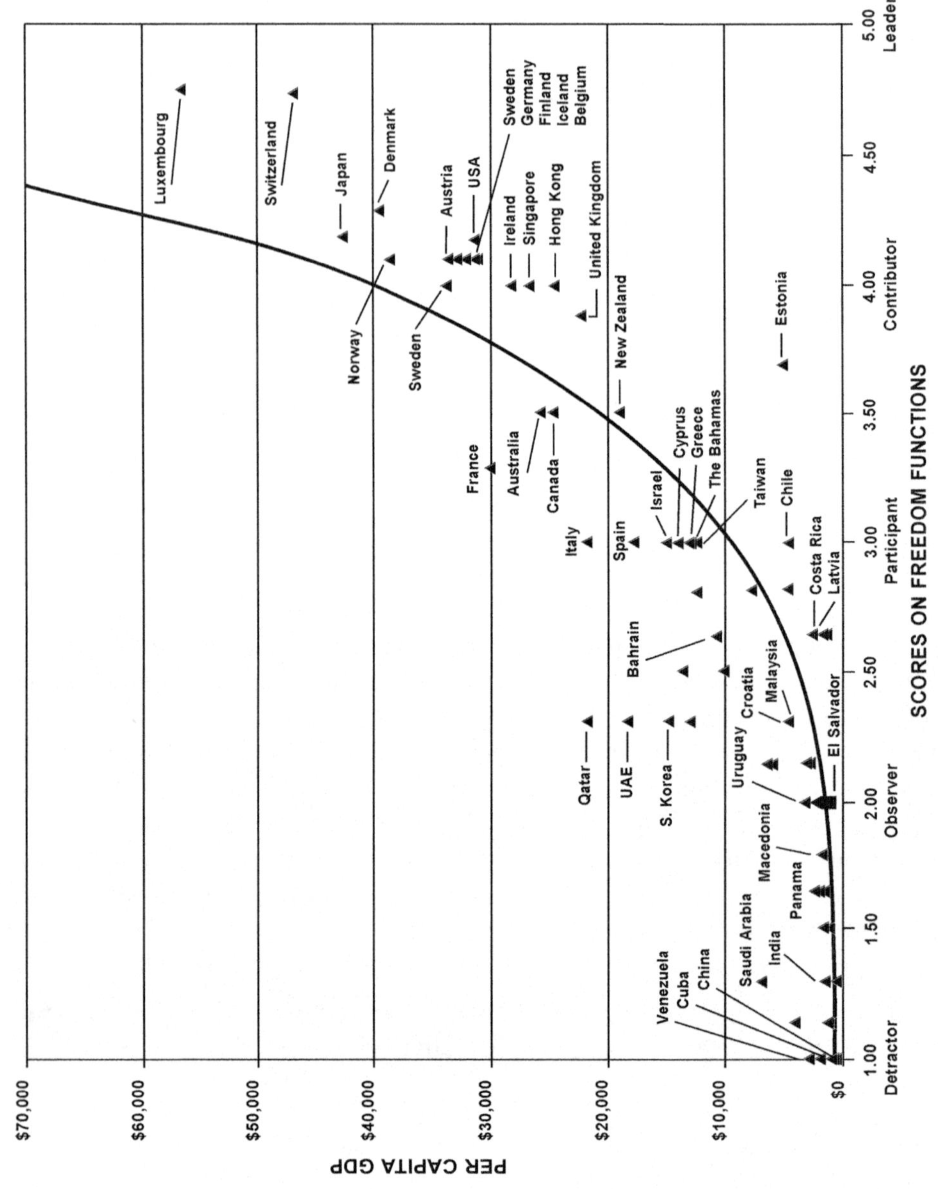

Figure 9. Freedom and Prosperity

In this context, Figure 9 presents the images of "the past, the present, and the future of civilization":

- The "past" is found in the clustered low scores on the freedom functions of "the detractors" who live in a version of "the pre-20th century civilization":
    - Dependent cultural relating
    - Totalitarian governance
    - Controlling economic enterprise
- The "transition from the past to the present" is discovered in the "middling scores" of "the observers" who live in a version of "the early 20th century civilization":
    - Competitive cultural relating
    - Authoritarian governance
    - Command economics
- The "present" is represented in the mixed scores of "the participants" who have only begun to embrace freedom in a version of "the mid-20th century civilization":
    - Independent cultural relating
    - Mixed governance
    - Mixed enterprise economics
- The "transition to the future" is found in the high scores of "the contributors" who are committed to becoming free and prosperous in a version of "the late 20th century civilization":
    - Collaborative cultural relating
    - Representative democratic governance
    - Capitalistic enterprise economics
- The "future" is modeled in the exemplary performance of "the leaders" who are attempting to commit fully to becoming free and prosperous in "the 21st century civilization":
    - Interdependent cultural relating
    - Direct democratic governance
    - Free enterprise economics

Indeed, all peoples of all nations may become free and prosperous. The freedom architecture is modeled by the exemplary performers on the freedom functions. This is the foundation for all possibilities civilization.

To be sure, there are no prosperous totalitarian nations!

Only nations striving toward freedom receive the rewards of freedom.

The threshold for prosperity appears to be at the participant level of the freedom functions (Level 3.0):

- Peoples and nations that relate with cultural independence, meaning collaborative or competitive according to the conditions and their requirements
- Peoples and nations that engage in mixed governance, meaning representative or authoritarian according to their circumstances
- Peoples and nations that engage in mixed enterprise economics, meaning capitalistic or by command as appropriate to their conditions

At the participant level, then, nations range in the $10,000–$20,000 per capita GDP.

Above the participant level, nations and peoples take off in growing and, then, spiraling levels of prosperity. At the contributor level (above Level 4.0), the modal earnings are in the range of $30,000 per capita GDP. Approaching the leader level (above Level 4.5), a nation such as Switzerland earns above $45,000 while Luxembourg reaches over $56,000 per capita GDP.

Below the participant level, nations are poor. At the observer level (above Level 2.0), where nations have not yet committed themselves to freedom, the per capita GDP ranges below $5,000. At the detractor level where many of the nations are reactive and even destructive, the earnings range in the hundreds of dollars.

This, indeed, may be the most important correlational figure of our time—or any time! Freedom is directly and powerfully related to generating prosperity and all of its related values—peace and participation!

In summary, we reap what we sow! Put another way, we get what we process for. If we process for freedom in collaborative cultural relating, free enterprise economics and participative democratic governance, we will achieve untold possibilities of prosperity and enlightenment.

If, on the other hand, we plan for control in totalitarian governance and command economics, we will achieve the unreported problems of isolation, poverty, and ignorance.

In transition, freedom assumes the infinite potential of human brainpower and empowers us to generate not only a continuously expanding economy but also our continuously changing destiny. Its adherents learn more and more about more and more until they know a great deal about an evolving-everything.

Totalitarianism assumes the finite nature of all resources from which derives an economy of scarcity, a psychology of stasis, and a pathology of discrimination. Its practitioners know less and less about less and less until, finally, they know nothing about anything.

Part III

Individual Generativity

5

Processing Systems

We can see the prototype of human processing in the first year of a child's life. The child learns to explore, understand, and act upon his or her world. What goes on during this first year of life continues in more and more refined ways throughout our lives—or not! How productively we live our lives depends totally on how effectively and efficiently we learn to process.

A great deal depends on how we develop our human resources. For example, it depends on how the environment—which is largely human—interacts with our biological selves. The newborn infant enters the world with few skills other than physiological reflex responses. For example, the child has the sucking and the palmar or grasping reflect. If the environment is responsive to the child, these reflexes will become instrumental in the child's survival. They constitute the child's initial movements toward the world that will ultimately lead to his or her growth and development. The child will be able to nurse with the sucking reflex. Later on, the child will be able to manipulate things with the grasping reflex.

In the beginning, however, newborn infants bring little but their inherent resources to their worlds. In their utter dependency, they wait for us to ensure their survival by responding to their needs and by gradually guiding them to the things they need to have to maintain themselves.

One of the ways that we guide our children is by helping them form habits. Basically, human habits are behaviors that are acquired without human intelligence. They can be acquired by associating or relating in space and time, two or more sets of activities. At least one of these activities must satisfy some human need in order for the behavior to be repeated as a habit. For example, the child may develop the sucking habit when nourished by the mother's breast. The results may be said to be instrumental in satisfying the child's need for nourishment.

In the process, the child may develop a "conditioned" sucking response to the stimulus of the mother's nipple. In a similar manner, the child may later on develop a conditioned grasping response to the stimulus of food, which is instrumental in satisfying the child's need for nourishment. There are many other kinds of life habits that can be developed without human intelligence or intentionality. The habits are "learned" only in the sense that they are repeated. They are not learned in the sense of being the product of human understanding. In fact, these habits are conditioned, spinal responses, not processed responses.

Human learning and, indeed, human intelligence begin to manifest themselves when children are several months old. At this point, children begin to explore themselves and their environment. They discover the existence of and the relationships between environmental stimuli and their own responses. In other words, children become aware of the association of **stimuli** to which they have become conditioned to and the **responses** that have been conditioned to the stimuli. They become aware of causes and effects in their worlds.

This awareness is a two-way street. For example, children become aware that the nipple or the food serves as stimulus to a sucking or grasping response. This response, in turn, will lead to satisfying a need for nourishment. Children may then become aware that a need for nourishment stimulates the response of search for the nipple or the food.

Through exploring, children become aware of both their past and present relationships to their environments—including themselves. Children attempt to describe where they are in relation to themselves and the world around them. **Exploring** is the first stage or phase of human learning. This form of exploration begins to distinguish humankind from all other forms of life.

It is a short step from becoming aware of the ingredients of human experience to anticipating experiences. With increasing confidence in this awareness of the relationship of stimulus and response, children are prepared for instrumental or purposeful learning at about one year of age. In other words, children set out to obtain certain results or ends. For example, children may set out to attract their mother, or to obtain food or an object that is out of reach.

Drawing from this awareness of the relationship between stimulus and response or cause and effect, children set goals of achieving certain effects. The goals of the instrumental act are often only seen later although some approximation of them was obviously intended from the beginning.

In short, children understand their relationships to future events or experiences. They are attempting to predict the consequences of their efforts. They understand where they want to be in their worlds. This **understanding** is the second stage or phase of human learning. It is what allows humankind to anticipate its future—which further distinguishes us from other forms of life.

The next phase of human learning flows naturally from the understanding phase. It involves the development of behavioral patterns instrumental to achieving goals. From the end of the first year onward, children draw from their repertoire of behaviors to produce the responses needed to achieve a goal. For example, children may laugh or cry to bring their caretakers to them. Children may move their hands in the direction of unreachable food or objects. There may be a series of trial and error experiences. These experiences may either confirm the child's responses because the goal is reached and he or she experienced satisfaction, or else the experiences may deny the child's responses because the goal is not reached.

Children begin to act in order to get from where they are to where they want to be within their worlds. They are attempting to control themselves and their worlds. **Acting** is the third stage or phase of learning. It enables human beings to plan and work toward the end of influencing their future.

Human Processing

The first year of human development serves as a prototype for all human processing. Children's reflexes are unknowingly conditioned as habitual responses to certain stimuli. These habits serve as the limited repertoire of responses with which children initially approach the world. Improvement in the quantity and quality of responses with which the learner ultimately relates to the world depends on the development of the children's intelligence. This, in turn, depends on how effectively they go through the stages or phases of learning.

Initially, children *explore* and identify the nature of the stimuli and responses in their experience. Then, children come to *understand* the interactive nature of stimuli and responses, anticipate the effects of one on the other, and develop goals to achieve these effects. Finally, children *act* by drawing from their developing repertoire of responses to attempt to achieve goals. The children's action behavior is shaped by feedback or by the effects it achieves. This feedback is recycled as input through the stages or phases of learning as the children explore more extensively, understand more accurately, and act more effectively. This ascending, enlarging spiral of exploration, understanding, and action is the source of the adult's improving repertoire of responses.

It is important to note that as learners develop, they may process differently depending on their own learning styles and functions. For example, some learners may *begin* with an action response that they later explore and understand. Nevertheless, the exploring–understanding–acting sequence may be cycled to produce more effective and efficient action responses.

The Skills of Exploring

Just watch the learners' exploratory activities in school. Perhaps we can see these in a classroom most clearly when there is an absence of formal teaching. The learners may become interested in some object or mechanism, on their own initiative, in the absence of a teacher. The learners will approach the material or object and position themselves so as to give it their attention. They may observe it for a while, perhaps listen to it, and then probably touch it. The touching will lead to handling. The learners may try this out in different ways, turning the material around or over, or attaching it to other things.

The learners may try to figure out what the thing is and does, and maybe even why and how it does it. Finally, they may try to do whatever it does. In the process, they have found out what they know about it and what they can do with it. In short, they have found out where they are in relation to the learning experience. (So have their teachers, if they are present.)

The learners must address all sources of learning in a similar manner. If the teacher is presenting content in the classroom, the learners must use all of their exploring skills to address the teacher, the content, the delivery or method, and the classroom environment as dimensions of the learning experience. Learners must also address themselves as potential sources of learning, in terms of the learning experiences and learning skills that they bring to the learning process. By exploring all dimensions of the learning experience, the learners can find out *where they are* in relation to the learning experience. They will then be ready to find out *where they want to be.*

The Skills of Understanding

The learners may engage in a series of understanding activities. They may relate the dimensions of their current experiences to past experiences in different ways based on their similarities and differences. The learners may organize the dimensions of the learning experience in still different ways based on their functions and the values of these functions to their learning. The learners may generalize their needs for these values and set generalized learning goals based on the learning experience and specific learning objectives derived from the different dimensions of the learning experience. Or they may do all of this by simply determining what, of all the content possible, they have yet to learn.

Where there is a teacher with a teaching goal, all of these activities can take place in relation to that goal: the learners may set their learning objectives in relation to the teaching goals. Where there is no teacher, the learners may set their learning objectives based on their generalized needs. In other words, the learners gain increasing confidence in their understanding of where they are in relation to where they want to be. They are ready to act in order to get there.

The Skills of Acting

Observing the learners in the action phase reveals the types of activities in which learners engage. First, they work to master the knowledge and skills involved in the learning goals. If the teachers have established the goal, they learn the knowledge or skills the teachers have developed. If teachers are not involved, learners can begin to define their skills objectives in terms of the deficits or problems they are having in achieving their goals and to develop and implement programs designed to achieve those objectives.

In either event, having acquired the learning, the learners can repeat or practice the skill involved until it is ready and effectively available to them. Then they can apply it, either in some way that was intended by the teacher, or that is relevant to their own experience. They can continue to apply it in real life, everyday living, learning, playing, and working experiences. Finally, they can transfer the learning to unique and creative situations in their lives. This is the culmination of mastery: to be able to use what you have learned to create.

The Phases of Human Processing

People in the everyday world process their experience in a similar manner. They process the information about people, data, and things they receive as input. They explore, understand, and act in order to transform these stimulus inputs into productive response outputs.

For example, parents in a living environment may explore, understand, and act on their children's behavior, information regarding their mortgage, or the installation of new equipment. Youth may process parents' or peers' values, information on their school or course options, and instructions regarding use of a personal computer.

Similarly, teachers in a learning environment may process students' learning behavior, data regarding students' learning performance, and information regarding the use of a new computerized data system. In turn, the learners may process information regarding peoples of other cultures and data regarding their own aptitudes and abilities, and learn to manage to manipulate the two extraordinary instruments of our time—the automobile and the computer. In this context, the compatibility of teacher and learner objectives is critical. *The productive teacher accommodates the learners' functional objectives* instead of allowing the learners to merely adjust to the teacher's processes.

Processing in the work setting is similar. Policymakers and managers process primarily information on people or data while supervisors, delivery personnel, and production personnel process raw materials into finished products, raw data into useful information, and naïve candidates into productive workers. In the working context, unlike the teaching/learning context, productive delivery personnel must incorporate the goals and objectives of their managers and supervisors in implementing their tasks. However, increasingly in the high tech and knowledge businesses, the tasks are shared and then merged interdependently. Over time, the middle and lower levels of management will disappear, as delivery personnel process all aspects of data with the purpose of improving self-management and performance.

6

Individual Generativity Processes

Generative processing has been the theme of my life. It is the source from which all breakthrough sciences are generated. It is the force from which all cutting edge technologies are innovated.

It is also the most difficult human experience to articulate: that the raw materials of human experience produce the finished products of human endeavors; that random information inputs are transformed into invaluable new capital outputs—individual, organizational, community, cultural, economic; that seemingly nothing is made into something.

Our mission is to transform IT "users" into "co-processors":

> **To prepare co-processors for IT programs by empowering them with human generativity systems that process the levels of information representation of the phenomena being addressed.**

This is the mission of Generativity Science for the "IT world":

Generativity Science ↔ (HC ↔ IC)

Generativity Science is interdependently a function of the synergistic processing relationship between **Human and Information Capital: (HC ↔ IC).** Each grows as the other grows!

Goaling by Relating

We initiate all of our generating processing experiences with goaling. In order to define my goals, we must relate to the conditions of my environment. So we must ask and answer the following question:

What are the dimensions of our processing environment?

Extensive factor analysis yields two prepotent factors upon which all other dimensions load:

1. Information Representation or IR
2. Human Processing or HP

Both factors need to be scaled in order to determine the effects of each level upon every level of both factors (see Figure 10).

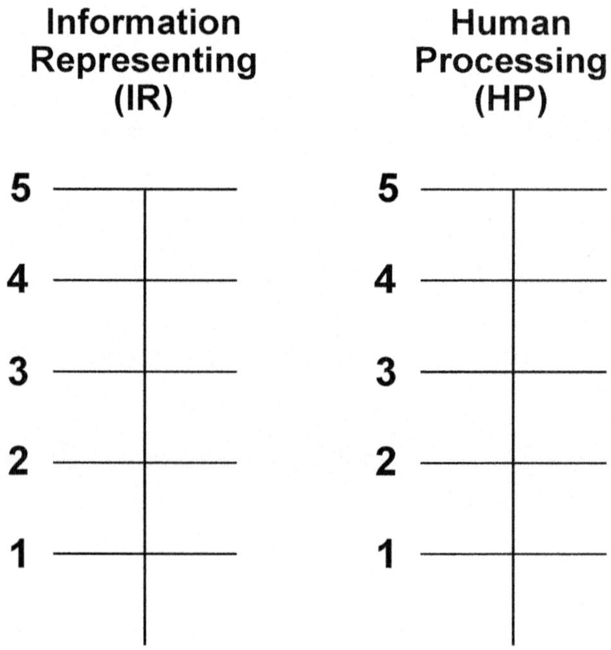

Figure 10. The Prepotent Factors of Generative Processing

Inputting by Representing

Now that we have the factors of generative processing, we need to relate them. We relate them inductively, as illustrated in Figure 11, where low levels of **HP** (1 2 3 4 5) are aligned with low levels of **IR** (1 2 3 4 5). We also may learn to relate them deductively by aligning high levels of **HP** (5 4 3 2 1) with high levels of **IR** (5 4 3 2 1). This matrix defines the boundaries of the **human universe** in which we are processing. This **Human Generativity Matrix** will tell us the effects of each level of dimensions upon every other level of both factors (see Figure 11).

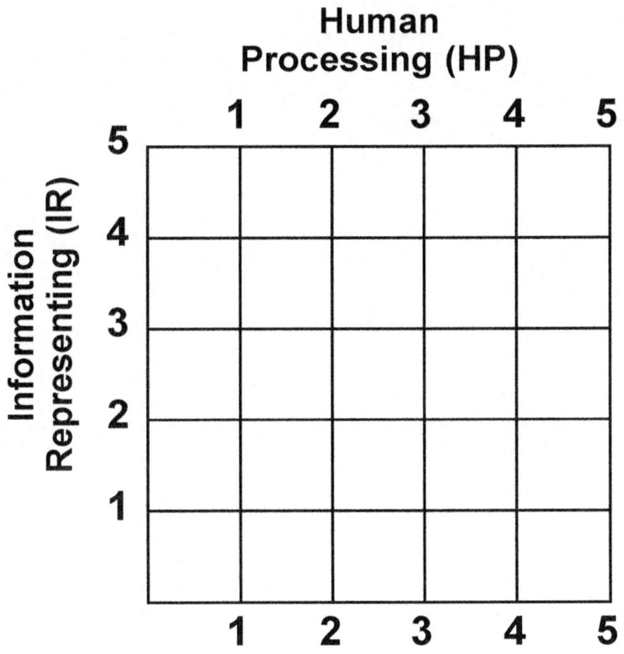

Figure 11. The Human Generativity Matrix (Inductive)

Exploring by Expanding

Now that we have defined the human universe, we can initiate **human processing** or **reasoning.** The first stage in **reasoning,** or **R[1]** is **exploring by expanding** (see Figure 12). As may be noted, expanding involves going out to explore and analyze the highest levels of **IR.** This gives us the greatest perspective on the lower levels of information representation. In other words, we go as high on **IR** as is necessary to understand the phenomena we are addressing.

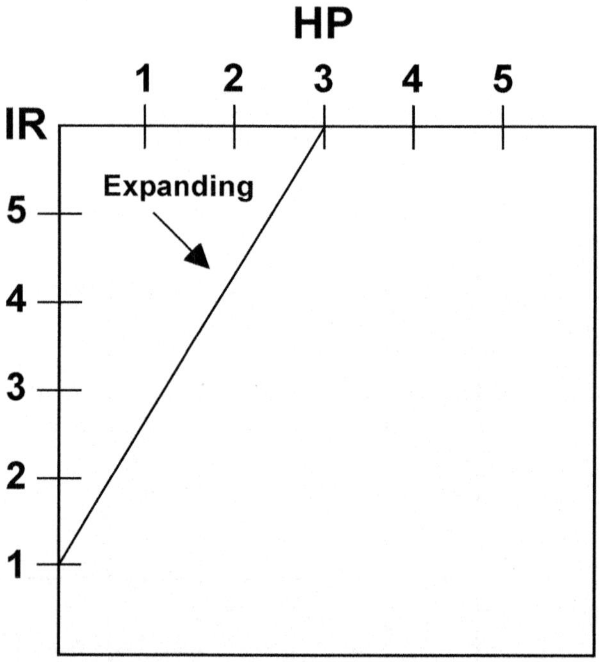

Figure 12. Exploring by Expanding

Understanding by Narrowing

Now that we have the greatest perspective, we seek to narrow our objectives by defining both our values and the conditional requirements. In other words, just as we view our universe with our values, so does our universe view us with its values: **"Do we meet its requirements?"** If so, we can go on to define the operations of our objectives (**O**) (see Figure 13).

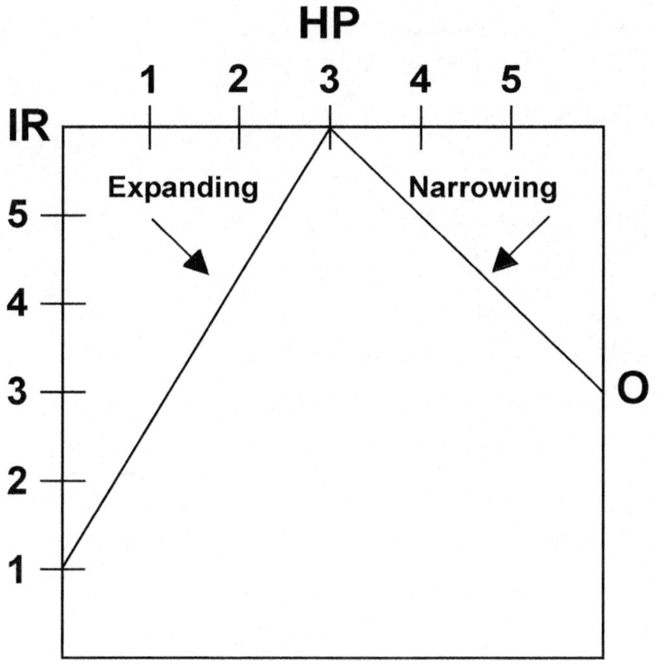

Figure 13. Understanding by Narrowing

Acting by Objectifying

Finally, having narrowed on our target objectives (**O**), we can now define them operationally—$\mathbf{O^1}$, $\mathbf{O^2}$, $\mathbf{O^n}$ (see Figure 14). In other words, we can define the **Os** in operational dimensions:

- **Functions:** Responses or Products
- **Components:** Parts or Participants
- **Processes:** Procedures or Methods
- **Conditions:** Contexts or Environment
- **Standards:** Achievement or Excellence

These operational definitions make our **Os** achievable.

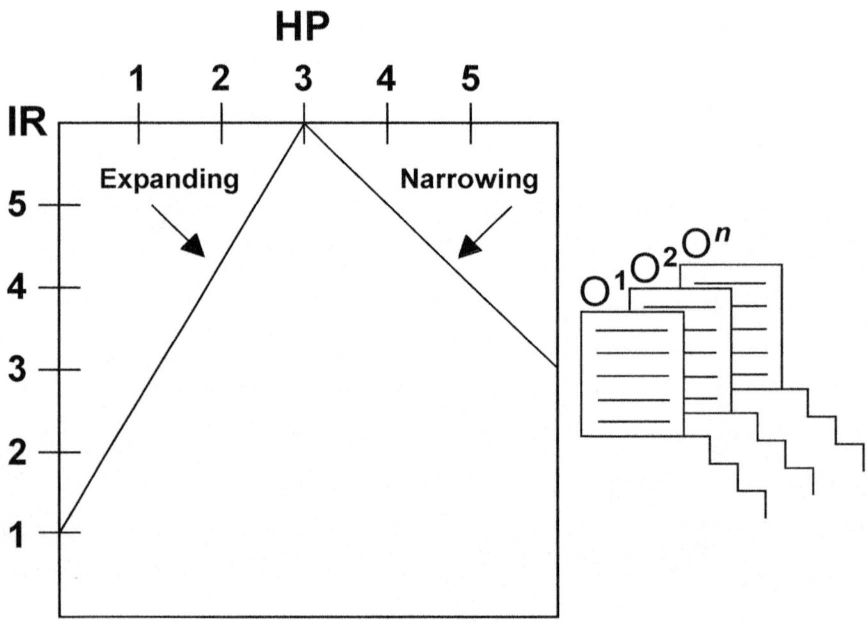

Figure 14. Acting by Objectifying

Generativity is **"thinking beyond the high beams."** It is as if we were driving along as we use our brainpower in most of our lives—using our low beams. Once in a while at night, we switch to high beams and see things that we would never have otherwise seen. This enables us to avoid accidents with the obstacles lying ahead of us.

When we flick on our "thinking beams," we see beyond the high beams. This empowers us to see not only things that might have been, but also to see ourselves in relation to things that might be—in other words, possibilities. This empowers us to view the possibilities in our lives. Even in the face of failure, there are exponential and sometimes infinite degrees of success.

We may take a simple overview of the structure and functions of **human processing** or **generativity thinking operations.** As will be seen, the phenomena is a simple one. It begins with a simple image of information input or knowledge or data:

It ends with a concrete image of information output, or action plan:

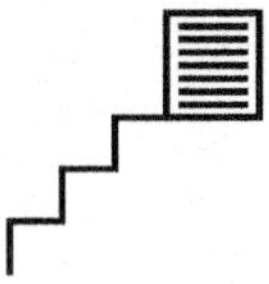

Generativity goes through two simple processing operations to transform the information input into action output: expanding and narrowing. First, we *expand* the possibilities of operations. Then, we *narrow* to the preferred probabilities operation that will define the objective of our action plan.

This is the simplicity of generativity:

- First, we *relate* to images of information.
- Second, we *represent* these images operationally.
- Third, we *expand* more possibilistic images.
- Fourth, we *narrow* to a preferred probabilistic image.
- Fifth, we *prepare* an action plan.

As may be noted, the generativity operations define a **generativity human processing system.**

The Individual Generativity Paradigm

We may view our representation of the paradigm for human processing in sharpest relief in the **New 3Rs** (see Figure 15). As you may see, the paradigm unfolds as a branching system or tree:

- **R^1 Relating**
  - Getting images
  - Giving images
  - Merging images
- **R^2 Representing**
  - S^1 Sentences or conceptual images
  - S^2 Systems or operational images
  - S^3 Schematics or dimensional images
- **R^3 Reasoning**
  - Exploring or expanding alternatives
  - Understanding or narrowing to preferred alternative
  - Acting or doing by performing tasks to achieve objectives

Note that the New 3Rs are the critical phases in systematically generating new and more powerful responses:

- Relating by sharing and merging
- Representing by operationalizing and dimensionalizing
- Reasoning by exploring, understanding, and acting on alternatives

In reverse, we cannot reason without expanding and narrowing alternatives: we cannot expand and narrow without representing images; we cannot represent images without relating to information.

This paradigm illustrates the power of **individual generativity:**

- It is operational.
- It is measurable.
- It is replicable.

In other words, it meets the criteria of **probabilities science** and **parametric measurement.**

It also meets the criteria of **possibilities science** and **non-parametric measurement:**

- It is developmental.
- It is cumulative.
- Its operations are not predetermined.

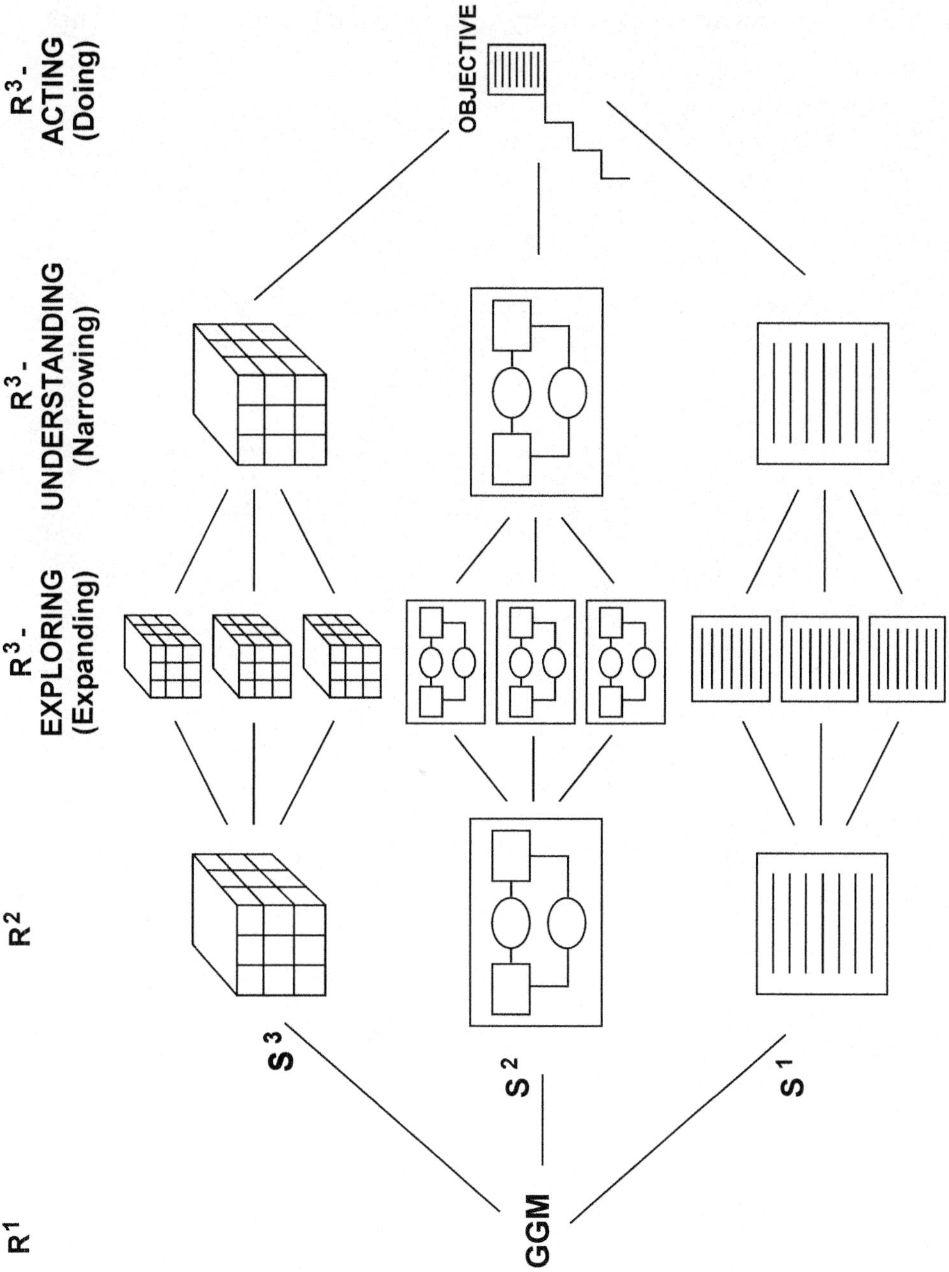

Figure 15. The Individual Generativity

---

*The generative human processing system is available for viewing in animation on Pod III of **mcleanproject.com.**

In other words, individual generativity searches out the truth wherever it lies.

- By exploring our experience fully.
- By understanding new goals accurately.
- By acting on new programs concretely.

7

Individual Generativity Outcomes

Our review of recent studies viewed the evolution from **"Probabilities Schools"** through **"Possibilities Schools"** to **"Generativity Schools."** Initially, the schools were dedicated to S–R conditioned responding functions. In other words, the students were directed to making the correct conceptual responses to specific stimulus items. Through our work in interpersonal relating skills, we were able to empower teachers to relate the learners to the conceptual information that enabled the correct responses to the stimuli. In other words, we intervened to make the Probabilities Schools effective by training the teachers in **relating** processes in order to achieve **S–R** conditioned responding functions.

With the introduction of the Data Age, we introduced Possibilities Schools. These schools were dedicated to S–O–R learning systems. In order to accomplish these student functions, they dedicated operational information components. Finally, in order to enable the operational components to discharge **S–O–R** learning functions, they complemented **relating** processes with **representing** processes. These were the transitional Possibilities Schools, emphasizing **S–O–R** discriminative learning functions.

Finally, with the spiraling requirements of the evolving Information Age, the Generativity Schools were introduced to fulfill their generativity science mission. These schools were driven by S–P–R generative processing functions: they were powered by dimensional information components enabled by **reasoning** processes. For educational purposes, we termed the learning processes **The New 3Rs: Relating, Representing, Reasoning.** They were the skills required of our learner-products in the 21st century global marketplace. They were the skills required to actualize a truly **"Human-Centric Information Age."** These were Generativity Schools emphasizing **S–P–R** individual generativity functions.

The results from these studies were illuminating and exhilarating.

- The Generativity Schools had the nation's top SAT scores.
- The Generativity Schools had over a 90% retention rate for hundreds of thousands of students in culturally diverse states.
- The Generativity Schools quadrupled the Advanced Placement enrollees.

All of this was accomplished while strengthening the career and educational programs as well as other options.

From a database of over 200,000 participants of human endeavors (see Table 2), we can hypothesize the following:

- Systematically trained helpers (parents, teachers, managers) have a 95% chance of being successful with helpees in all kinds of human endeavors: living, learning, and working functions.
- Systematically trained helpees (children, students, workers) have a 95% chance of being successful in all kinds of human endeavors.

Banks and associates have summarized the human meaning of these results with the skills of individual generativity:

- **R^1 – Relating** facilitates **entry** with goal-setting.
- **R^2 – Representing** facilitates **transition** with operations.
- **R^3 – Reasoning** facilitates **exploitation** by analyzing.
- **R^4 – Reasoning** facilitates **understanding** by synthesizing.
- **R^5 – Reasoning** facilitates **acting** by operationalizing.

Relatedly, Banks has summarized the effects of **"The New 3Rs"** in **The Generativity Schools** he has studied upon learner achievement of Standards of Learning:

> "In **Probabilities Schools,** we have demonstrated that educators can be taught to strengthen and use instructional processes (curriculum development, lesson planning, teaching delivery) with relation to **'Old 3Rs' Information Components,** to increase student achievement on state Standards of Learning. For the subjects of math, English, social studies/history, and science, student achievement has increased approximately 140% over three academic years. These academic achievement gains were for both white and minority students.
>
> "In **The Generativity Schools,** student achievement on **Standards of Learning** were increased as educators learned and used the **'The New 3Rs' Instructional Processes** with relation to **'Old 3Rs' Information Components.** Student achievement in high school English (reading and writing) reached a student pass rate of 90% to 95% for both white and African American students, illustrating a significant reduction in the gap between higher levels of white student achievement and lower levels of minority student achievement.*"

Studies cumulating in the largest database on human endeavors appear in Table 2.

---

* Banks, G., Banks, K., et al. *The Effects of "The New 3Rs" Upon Learning "The Old 3Rs."* McLean, VA: Carkhuff Thinking Systems, 2012.

Table 2.
The Effects of Relating and Other Skills of Individual Generativity

OUTCOMES	HELPERS	HELPEES	OUTCOMES
LIVING			**LIVING**
Studies (N = 22)	91% Positive	91% Positive	Studies (N = 35)
Indices (N = 117)	83% Positive	84% Positive	Indices (N = 128)
LEARNING			**LEARNING**
Studies (N = 32)	97% Positive	100% Positive	Studies (N = 26)
Indices (N = 261)	92% Positive	99% Positive	Indices (N = 78)
WORKING			**WORKING**
Studies (N = 22)	100% Positive	100% Positive	Studies (N = 27)
Indices (N = 83)	96% Positive	98% Positive	Indices (N = 117)
SUBTOTAL			**SUBTOTAL**
Studies (N = 76)	97% Positive	96% Positive	Studies (N = 88)
Indices (N = 461)	92% Positive	92% Positive	Indices (N = 323)
GRAND TOTAL			
Studies (N = 164)	96% Positive		
Indices (N = 784)	92% Positive		

In short, Banks concludes that the simple, most important set of skills that anyone can teach the next generation of learners is generative thinking or processing skills:

> "It is the **'Holy Grail' of Education.** It is the 'great equalizer' of conditions. It neutralizes the effects of all demographics—it empowers the learners to process all content and make it instrumental for their own purposes."

Generativity processing is the theme of the book: **To learn to think in generativity terms.**

Part IV

Organizational Generativity

8

The Greatest Story Never Told*

This is the greatest corporate story never told. The effective ingredients of this "fly-over" are known only to a select few people: those who conceptualized and then operationalized the powerful interventions that led to changes in Airbus' policies and practices. Indeed, the corporate officers never realized that they had a tiger in the tank.

Airbus, a consortium of European aerospace industries, only learned to fly in 1970. Caught in the draft of Boeing, it faced financial-survival flying conditions until the 1990s, because Boeing dominated commercial aircraft production with a 70% market share.

These were the market conditions when I, as Executive Vice President of Marketing, Computervision, offered state-of-the-art Electronic Product Definition (EPD) services: that is to say Computer-Aided Design, Analysis, Manufacturing, and Product Data Management. Unbeknownst to us at the time, we were also offering Airbus a ticket to its flying future.

Enter Bob Carkhuff and his associates at Carkhuff Thinking Systems, Inc. Carkhuff and his associates were working with new and powerful ideational constructs in the economic arena. They labeled the core construct "New Capital Development" or NCD. Reasoning that financial capital accounted for approximately 10% or less of economic productivity growth (necessary but not sufficient), they set out to demonstrate the potency of these NCD ingredients as the sources of organizational generativity (see Figure 16):

- **MCD** or Marketplace Positioning
- **OCD** or Organizational Alignment
- **HCD** or Human Processing
- **ICD** or Information Modeling
- **mCD** or Mechanical Tooling

They argued that together, these NCD ingredients could account for the greatest part of economic growth.

* Barry Cohen, Ph.D., Vice President, Parametric Technology Corporation

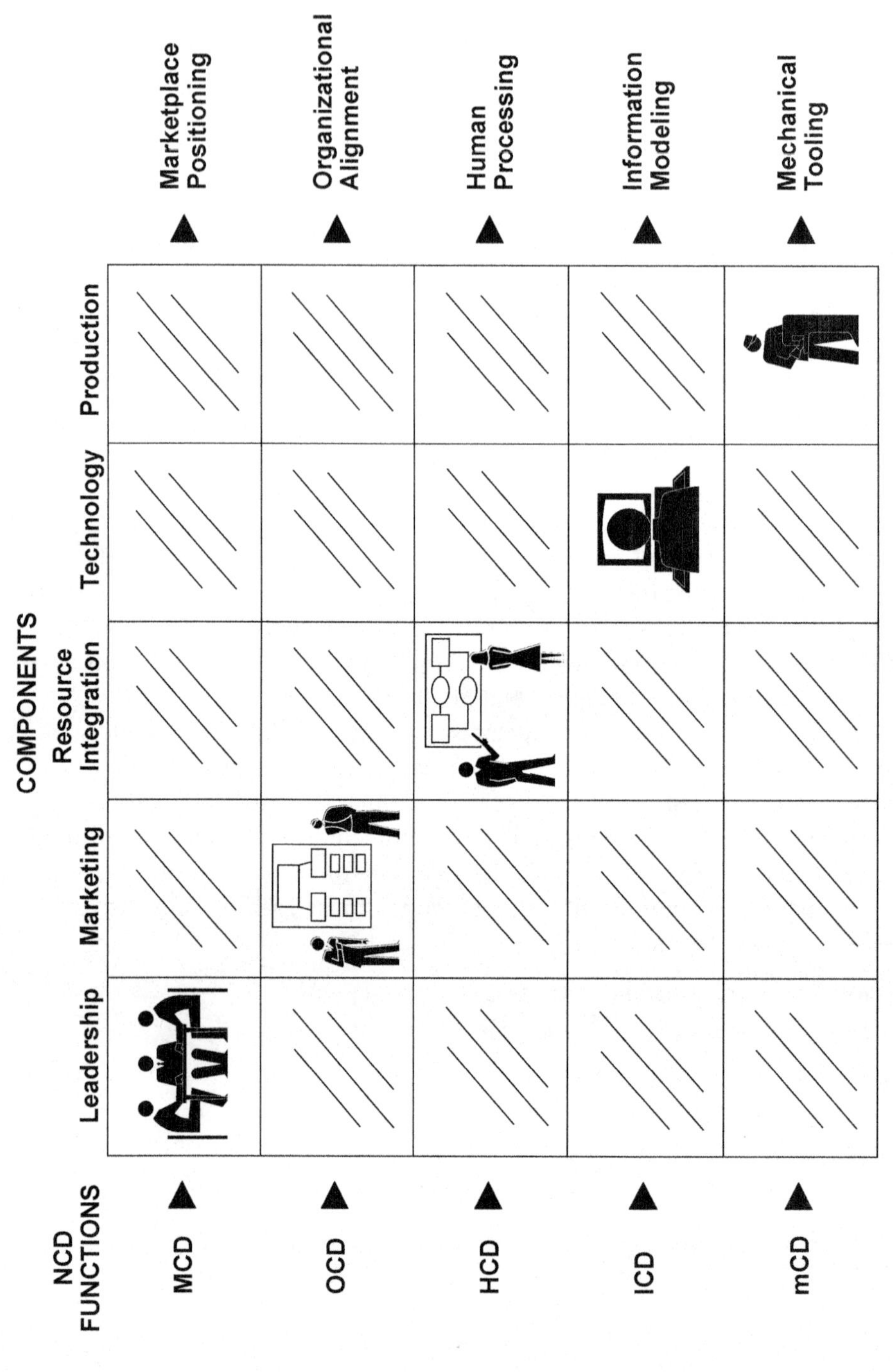

Figure 16. Organizational Generativity

Now we were set for the greatest experiment in corporate history. The ideational intervention that would change the Airbus policy-makers' ways of thinking would begin with the NCD ingredients. Although the corporate board embraced the ingredients of NCD, its members never heard the term "NCD"!

Carkhuff Thinking Systems had supplied the appropriate NCD operations to Computervision, which at the time was processing with the marketplace to determine what responses it would need to make for growth. I remember vividly my initiation to the NCD concepts and constructs that would change forever the way I thought about human and organizational systems design and operation.

Our NCD processing took the form of intensive dialogues with our aerospace customers. The critical challenge faced by our new customer was how to participate in an aircraft design program that was being defined by 3D Computer-Aided Design (CAD), using digital solid models for the whole aircraft definition.

This was a very big deal, as it fundamentally changed the organizational processes between companies from independent cooperation (exchange of 2D drawings) to interdependent collaboration (shared 3D design space). Using NCD thinking, it was clear that the marketplace requirements had changed, in part, due to time-compression and cost-reduction constraints. This had triggered a new organizational process response that in turn demanded new human learning, changes in behavior, and new skills.

People would need to work with new Information Capital (in this case, virtual assembly spaces of 3D solid models) which then required a technology tool-stack of CAD authoring applications, product data management, hardware platforms and high performance IP networks. The benefit to be derived for all this change was a massive reduction in time to design the product and a very large cost reduction to be achieved by removing the need for multiple expensive physical prototypes. Electronic Product Definition (EPD) and Digital Pre-Assembly (DPA—later renamed IPD or Integrated Product Definition) were born in the heat of spiraling change, true Information Capital Development (ICD)!

However, the greater challenges faced were the responses to be made in the Organizational Capital Development (OCD) and Human Capital Development (HCD) domains of NCD. Given the changing conditions now faced, what were the new skills, learnings, and behaviors that people needed to adopt, and how quickly should this take place? The answers to these questions were nested in the context of having to know the organizational processes, project configurations, and workflows that were to be operated to deliver the products and services to market.

Computervision, a technology company that successfully innovated in response to the marketplace with EPD and IPD, could now commercialize on the Information Capital Development it had realized. However, during the innovation process, it also had developed and acquired new embryonic capabilities in OCD and HCD that were later labeled IP^5D or Integrated Process Development. These are methods by which organizational processes can be defined, dimensionally represented, and related to the human behavior, skills, and learning required for the successful adoption of new ICD and adaptation to new OCD.

Between these two domains is the domain of Human Capital Development, which becomes the critical success factor for any enterprise venture. The speed at which old learning can be unlearned and new learning can be deployed represents the social adaptive latency coefficient of an enterprise. Just-in-Time (JIT) learning, innovative thinking models, coaching, and mentoring were developed to facilitate and support enterprises in responding to change. Computervision had now developed powerful professional services to complement its technology solutions, and began to leverage these new responses, recommendations, and expertise in its marketplace; that success later led to its acquisition by Parametric Technology Corporation Inc. (PTC).

This, then, was how I came to be empowered with these NCD ideas and presented them in the form of EPD to the corporate board at Airbus. The reality was that Airbus was already conceptually NCD-centric, but not operationally. The company continuously processed with the marketplace, organized itself collaboratively and systematically, and reveled in the diversity of ideas generated. Airbus also invested heavily in Human Capital Development across different cultures, but lacked common Information Capital Development (ICD) and Mechanical Capital Development (mCD) tools, until the Computervision intervention.

Indeed, NCD was facilitated at Airbus by Computervision in the form of evolutions of ICD, driven by the NCD recognition that value-added market solutions comprise Organizational Processes, People, and Technology. These were defined at Computervision as:

- EPD or Electronic Product Definition that Computervision contributed;
- IPD or Integrated Product Definition that Carkhuff Thinking Systems contributed;
- IP^5D or Integrated Process Development that Carkhuff refined.

To be sure, Carkhuff's IP^5D represented the processing systems of NCD:

- **P^5**–Positioning or MCD Positioning;
- **P^4**–Partnering or OCD Alignment;
- **P^3**–People or HCD Processing;
- **P^2**–Processing or ICD Modeling;
- **P^1**–Products or mCD Tooling.

In other words, we at Computervision had embraced NCD and subsequently integrated NCD into the policies and practices of Airbus by our collaborative processing with the ACE (Airbus Concurrent Engineering) program. Enough of the New Capital Development orientation, processes, and practices were adopted by Airbus to help power its amazing growth over Boeing.

What follows is a composite, adapted story of Airbus vs. Boeing in a battle for market share in the commercial airliner production business. It is a story of the introduction, implementation, and emphasis of a corporation based on New Capital Development. NCD was the new ingredient for Airbus that powered its meteoric market assertion.

I document this as a personal history. More importantly, this story previews the story of the future of all business economic growth by New Capital Development!

Great corporate battles were fought by two megalithic airplane manufacturers: Boeing and Airbus. Initially, neither understood the real terms of the battles nor the true mission of the war. Conditioning dominated as each fought for the competitive edge in the marketplace. Then, marketplace conditions changed, and with them the directions of these two behemoths. Let us now take a *virtual look* into the two boardrooms as they anticipated these changes.

The Boeing Boardroom

Their marketplace positioning established as the premier builder of aircraft, the leaders of Boeing opted for a policy of incrementalism to deal with production bottlenecks and declining profits. The broad outline of the policy emphasized monitoring the computer systems, trimming thousands of jobs, and remaking relationships with the company's huge supply chain.

A redesign of assembly operations had actually been underway for several years. However, it got detoured by a big production push to compete with Airbus. Boeing won the competition, but was losing the edge: production costs spiraled and profits tumbled. Moreover, global economic problems sent Boeing into disarray, leaving many orders unfilled.

Nevertheless, Boeing had a lot more going for it than met the eye on the profit sheet. In addition to its historic positioning, it had futuristic positioning as exclusive builders of projected products such as U.S.A. multi-services military aircraft. Indeed, it had already acquired McDonnell Douglas and its proprietary technologies in preparation for these bidding battles. Moreover, it had a massive technology transfer project with the emerging Chinese nation in which Boeing would exchange turnkey empowerment for continuing orders of aircraft, 30 being the latest and greatest number.

Then, surprisingly, Boeing received an invitation from Airbus centered upon joint-venturing between these two largest competitors. These discussions would focus upon jointly building the world's best, cutting-edge products.

"Send them our bureaucrats!" was the order that emanated forth from the corporate boardroom.

"But they can only say 'No,'" was the reply.

"Exactly!"

The Airbus Boardroom

Airbus was positioned in the draft of Boeing's leadership. Indeed, Airbus often ate the scraps from Boeing's table—projects that were, for example, both mechanically and financially problematic. In other terms, Airbus got to customize or even tailor products that Boeing could not easily and profitably standardize.

In this regard, Airbus had developed an expertise in tailoring products to customer requirements. Whereas Boeing's value system revolved around conservative, probabilistic principles, Airbus' value system revolved increasingly around progressive, possibilistic principles.

In short, Boeing was disposed to stick to the knitting, avoiding risks, and protecting the status quo. Airbus, in turn, was increasingly disposed to being the best it could be, to taking risks, and to embracing change.

It was in this context that Airbus extended an invitation for joint venturing. Airbus had, itself, been a collaboration of efforts and specialties from the major aerospace businesses of Europe: namely, English wings, German fuselage, French cockpit and assembly, Spanish tailpane. Collaboration was a reflexive, albeit difficult process, and so the invitation was extended.

"Together, we can be the best in the world for as far into the future as we can see," was the rallying cry.

"Collaboration, not competition!"

"Interdependency, not independency!"

Unfortunately, it was not to be! With great hopes and much joy, the Airbus negotiators moved forward, only to be denied by the nullification of the Boeing bureaucrats! These were the first shots fired in what has become the enduring war of the ages: a war between status quo and change; a war between the old financial capital-driven economy and the New Capital-driven economy.

The NCD Wars

In this first foray, we can view the extremes of battle formations:

- Boeing's commercialism is dedicated historically to the potentially finite variations of legacy learning and commoditized applications.
- Airbus' commerciality is dedicated futuristically to the potentially infinite variations of generative thinking and innovative applications.

Both corporations compete on the same playing field:

- Boeing with a conservative and independent game plan, maximizing the contributions of powerful, existing technologies
- Airbus with a generative and interdependent game plan, emphasizing a commitment to developing all forms of new technologies

Neither corporation understood the mission of the war nor the terms of the battles. The mission pitted Old Capital versus New Capital. Old Capital meant the Financial Capital and Machine Capital that had dominated the equations of economists through the ages. New Capital meant the emerging new ingredients in the equation for economic growth.

Utilizing the exact same market data, then, the companies positioned themselves differentially in the marketplace:

- Boeing decided to go independent and continue on its then-current course, competing by customizing its standard product lines.
- Airbus decided to go interdependent and tailor future product lines to continuously changing marketplace requirements.

One corporation, Airbus, learned to process deductively from the marketplace mission:

- **MCD:** Positioning itself as an innovator
- **OCD:** Aligning its resources systematically
- **HCD:** Tailoring new products generatively
- **ICD:** Modeling information schematically
- **mCD:** Tooling production contingently

The second corporation, Boeing, grouped inductively to maintain market performance:

- **mCD:** Standardizing its old products
- **ICD:** Customizing new variations
- **HCD:** Driving learnings from legacy
- **OCD:** Aligning resources linearly
- **MCD:** Positioning itself as commoditizer

Today, Airbus is changing the innovative threshold in commercial aircraft manufacturing by introducing elevated standards for New Capital Development, upon which the current market is based. It is the pace-setter in the marketplace.

Boeing, in turn, is saved from attenuation only by the political machinations that make its commoditized products saleable on government contracts—here and abroad. Once a private sector standard-setter, it is now an extension of public sector bureaucracies.

Airbus has a future as innovator in the marketplace, based upon its growing treasury of NCD.

Boeing's future is tied to the rise and fall of non-substantive, political entities, based upon its legacy of Old Capital.

Airbus no longer seeks scraps from Boeing's table. Airbus seeks Boeing's lunch!

The Air Wars continue. Corporations come into and go out of existence, based upon their abilities to address continuously evolving and spiraling standards.

How do corporations win and lose wars so quickly? The answer lies in NCD—the New Capital Development ingredients of economic growth. The answer lies in understanding that our economic growth is a function of one form or another—or all forms together—of NCD. The Air Wars are really only the first battle in the "NCD Wars."

In summary, there really is a New Economy undergirding economic growth. We label it "The Generativity Economy." Indeed, one of the very reasons that recessions are shallow is the continuous elevation of Economic Productivity Growth or EPG. The sources of this EPG are the most important considerations in developing the Generativity Economy.

The current financial meaning of the term "capital" notwithstanding, the New Capital Development or NCD ingredients are now the prepotent sources of organizational generativity:

- **MCD** or Marketplace Positioning
- **OCD** or Organizational Alignment
- **HCD** or Human Processing
- **ICD** or Information Modeling
- **mCD** or Mechanical Tooling

Together, these NCD ingredients have displaced financial capital as the most important conditions of EPG. To be sure, financial capital remains a necessary catalyst to growth, accounting for somewhere less than 10% of EPG.

Together, these emerging NCD ingredients constitute new and expanding economic universes, and are the prepotent sources of spiraling economic growth. It is as if we are merely in the foothills of a volcanic and exploding mountain range created by the global market, the Internet, and other new platforms of free enterprise.

9

The ODNA Model

The organizational formula that has proved most leveraged is **"ODNA,"** the model for organizational generativity. We label it a *formula* rather than an equation because nature does not generate **equalities of operations.** Instead, we represent the interdependency of operations as follows:

ODNA ↔ *n*! (nest! encode! rotate!)

Labeled **"ODNA,"** we have assumed **DNA** or **Deoxyribonucleic Acid*** as our transfer paradigm for modeling all human functions:

- Attributing the highest **"quantitative"** measures to the most powerful **"qualitative"** levels of all human generativity phenomena as ***n*!** or **n factorial!**
- Hypothesizing the function of **"factorializing"** as **"nesting"** or **"housing"** by higher-order processing phenomena of lower-order processing phenomena
- Hypothesizing the function of **"nesting"** as **"encoding"** or **"programming"** of lower-order processing phenomena by higher-order processing phenomena
- Hypothesizing the functions of **"encoding"** as **"rotating"** or changing positioning in the various phenomenal systems

* DNA is the hereditary material containing genetic instructions for the development of all known living organisms.

[1] The Perspective of Einstein

There it was! Our formula for functional organizational systems:

ODNA → *n*! (nest! encode! rotate!)

This may be read as follows:

> **Organizational generativity** is a function of *n* dimensions **"factorialized" (*n*!)** with all dimensions **"nested," "encoded,"** and **"rotated."**

This is our guiding formula for the operations of organizational generativity.

It was the fall of 1988 and we were boldly employing our newly minted **"possibilities sciences"** to generate a model for **Organizational DNA.** Working together with my son, Chris, we had engaged all of our mentors to assist in building the **ODNA** paradigm: Bernie Berenson, our lifelong processing partner; B. R. Bugelski, our science and learning theory mentor; Jim Drasgow, our mentor for higher math and higher-order processing.

This was the formula that we were able to put together from processing interdependently with our colleagues. It was to lead to a powerful understanding of the universes of possibilities. Hopefully, over the course of this brief treatise, this will be made known to you.

Most of all, we were motivated by the fearlessness of Albert Einstein. Above all else, he taught us to go as high as we could go to get the greatest qualitative perspective on the phenomena, as he had done with energy and other phenomena of the physical sciences.

What gave our colleagues great confidence was their understanding of Albert Einstein's theoretical equations reflecting the relationships of man and nature:

$$E = mc^2$$

Einstein then made a great intellectual leap in transforming the algebraic equation: $\mathbf{E = mc^2}$.

In his second **Relativity** paper, Einstein made a historic deduction: If a body gives off an amount of energy (**E**) in the form of light, its mass will be reduced by that amount divided by the speed of light squared ($\mathbf{m = E/c^2}$).

This new formula established that even a small amount of matter held the power of tons of explosives. This opened the door to the Nuclear Age. It also led to explanations for why the sun could burn for so many billions of years while not shrinking appreciably in size.

Einstein himself said:

> It is customary to express the equivalence of mass and energy (though somewhat inexactly) by the formula $\mathbf{E = mc^2}$, in which **c** represents the velocity of light, about 186,000 miles per second; **E** is the energy that is contained in a stationary body; **m** is its mass. The energy that belongs to the mass **m** is equal to this mass, multiplied by the square of the enormous speed of light—which is to say, a vast amount of energy for every unit of mass.
>
> But if every gram of material contains this tremendous energy, why did it go so long unnoticed? The answer is simple enough: so long as none of the energy is given off externally, it cannot be observed. It is as though a man who is fabulously rich would never spend or give away a cent; no one could tell how rich he was (Einstein 1956, pp. 51–52).

So it is with human beings. All are fabulously rich with potential. Few ever spend or give away a cent.

[2] The Processing Paradigm

One of the most challenging projects that we encountered was IBM's Advanced Systems Development project on career development. Headed by John T. Kelly and Ted Friel, IBM failed miserably in attempting to validate career information dissemination for its *Education and Career Exploration Systems* or *ECES.* The dissemination of information by its computerized *ECES* system simply had no positive effect on its users. Indeed, dissemination may have constituted a **depressor variable:** the more information loaded on the user, the less likely the user was to use it to make career decisions or prepare for a career.

We simply asked the implementers **"Did you teach the kids to process the material?"**

We elaborated:

"Consider yourselves to have tested Condition I.

"In Condition II, we will teach the users to process the information."

The design for *ECES II* was revolutionary:

1. The users submitted their scaled career values.
2. The jobs represented their scaled requirements.
3. The user values and job requirements were interrelated in a matrix in the *ECES II* software design.

The results were astounding. Just as we had predicted, over 95% of the users were successful in career performance based on their *ECES II* experience. Simply because the user was put back in the career equation! Simply because the user was empowered to process!

In so doing, we designed the first interactive software system: processing human values and information requirements. Yet it was not to culminate during IT's domination!

We concluded that the **IT Revolution** was dedicated to replacing the human in the equation for human performance: **IT–IT–IT!**

In contrast, we have spent our lives developing a **"Human Technology"** or **HT** in the human performance equation: **HT–IT–HT–IT–HT!**

Our subsequent work was to "nest" or "house" the lower-order processing systems within the higher-order processing systems. In this manner, each lower-order processing system empowered the higher-order systems and enabled them to discharge higher-level functions. In turn, each lower-order system was encoded with the **"sociogenetic"** programs of the higher-order systems.

S–R Conditioned Responding

We may view the processing responses most clearly in our human processing paradigm. The most basic of processing systems are the **S–R Conditioned Responding Systems** theorized initially by John Watson and Ivan Pavlov. For humans, conditioned responding is a reflex, or knee-jerk, response. For other entities, it represents mechanical or natural processing. It is assumed that there is no organic or organismic processing between the presentation of the stimulus (**S**) and the conditioned response (**R**). However, in reality, there is the phenomenon of **chaining,** or anticipatory s–r responses (see Figure 17).

S ⟶ s - r - s - r ⟶ R

Figure 17. S–R Conditioned Responding Systems

S–O–R Discriminative Learning

Building on **S–R** conditioning, **S–O–R Discriminative Learning Systems** are the products of the generativity of Clark Hull, O. Hobart Mowrer, and B. R. Bugelski. Discriminative learning is a processing system that involves the organism (**O**) in intervening between the presentation of the stimulus (**S**) and the emittance of the response (**R**). The organism incorporates or *nests* a hierarchy of **S–R** responses with which to do the following: discriminate the uniqueness of the stimulus, select the appropriate response in the hierarchy, and emit the selected response (see Figure 18).

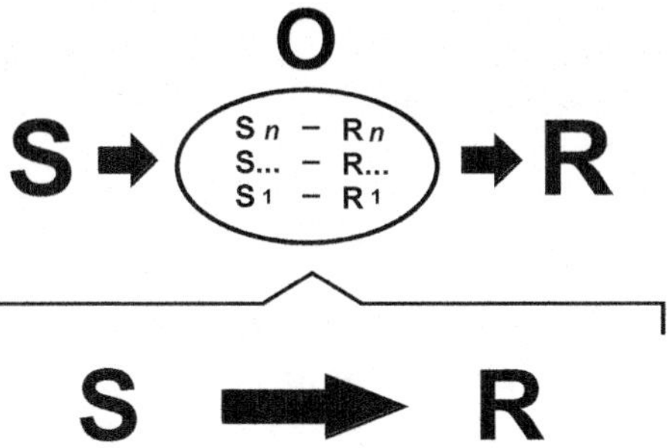

Figure 18. S–O–R Discriminative Learning Systems

S–P–R Generative Thinking

Building on the **S–O–R** systems of Hull and others, Carkhuff and associates generated the **S–P–R Generative Thinking Systems.** Generative processing is a higher-order processing system. It empowers the organism as a processor (**P**) with a hierarchy of discriminative learning (**S–O–R**) responses. This enables the processor to generate entirely new responses (**R**) that the stimuli (**S**) were not intended to elicit. In a continuously changing stimulus environment, the processor can generate a continuous flow of new and powerful responses to meet changing stimulus requirements (see Figure 19).

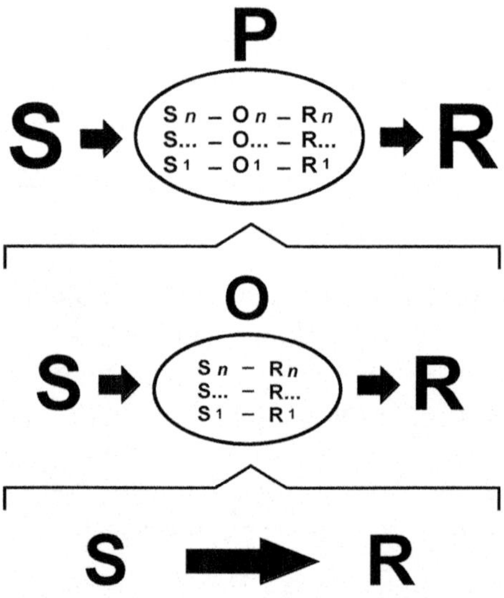

Figure 19. S–P–R Generative Thinking Systems

S–OP–R Organizational Processing

Higher-order processing systems such as organizational processing systems were our research mission at hand. Higher-order processing systems involve aligning our generative processing systems with the phenomenal processing systems. In order to do so, we need to discriminate the phenomenal processing systems. If we cannot discern these systems, we may intervene to empower the phenomenal processing systems. For example, human processors most often function within organizational processing systems.

S–OP–R Organizational Processing Systems are higher-order applications of generative processing systems. They empower the organization or organizational processors (**OP**) with hierarchies of generative processing (**S–P–R**) responses. In continuously changing organizational environments, the organizational processors can generate continuous flows of new and powerful responses to meet changing stimulus requirements: designing the organization, structuring its processing systems, aligning its resource systems (see Figure 20).

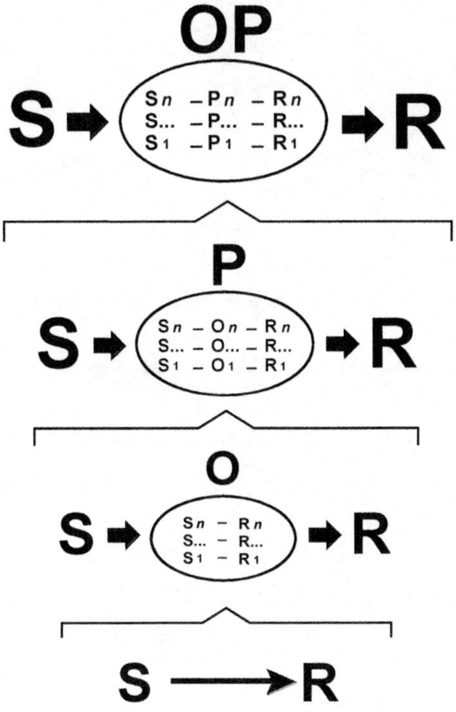

Figure 20. S–OP–R Organizational Processing Systems

S–PP–R Phenomenal Processing

For our purposes, then, we may posit higher-order phenomenal processing systems. For example, we may define marketplace or environmental processing systems as applications of generative processing systems. Phenomenal processing (**PP**) systems empower marketplace or environmental or universal processors with hierarchies of organizational processing (**S–OP–R**) responses. In continuously changing phenomenal environments, these phenomenal processors can generate continuous flows of new and powerful responses to meet changing stimulus requirements (see Figure 21).

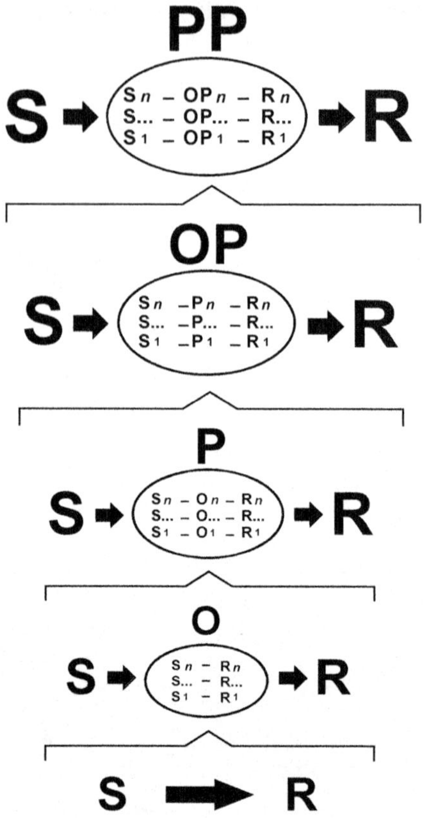

Figure 21. S–PP–R Phenomenal Processing Systems

The Processing Sciences

To summarize, the source of the **Processing Sciences** is found in the power of processing. Each level of processing is exponentially more powerful than the previous level—qualitatively as well as quantitatively, culminating in the highest levels of generativity processing. Those phenomena that are most empowered to process are most free to generate (see Figure 22).

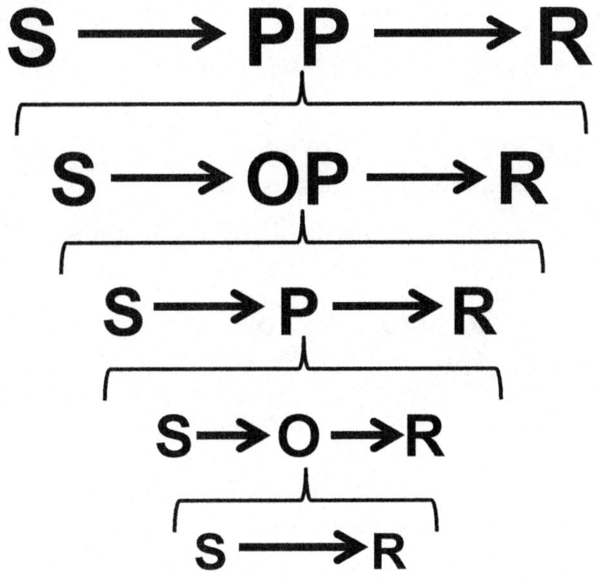

Figure 22. The Processing Sciences

All of this led B. R. Bugelski to comment:

> "This work rationalizes all of our efforts in learning theory and promises the culmination of psychology in a science of processing."

[3] The Organizational Processing Paradigm

The Processing Sciences provide the paradigm or model for the formulation of **Organizational Processing Systems** or **OPS**.

- The highest-order quantitatively (***n*!**) is qualitatively the most powerful processing system in the hierarchy of systems.
- The higher-order processing systems "**nest**" the lower-order processing systems.
- The higher-order processing systems "**encode**" the lower-order processing systems.
- All processing systems may be "**rotated**" inductively or deductively toward changing positions in the phenomenal systems.

There are five **OPS** steps involved in actualizing **ODNA Systems.**

OPS 1. Converting Organizations to Phenomenal Processing

The first step in utilizing this processing paradigm as the model for all human phenomena is to convert to the phenomenal processing systems. In this instance, the Organizational Processing Systems or OPS emphasize systems processing: Goaling, Inputting, Processing, Planning, Outputting. Each of these levels of OPS may be operationalized by the levels of **Phenomenal Processing Systems** or **PPS** (see Figure 23):

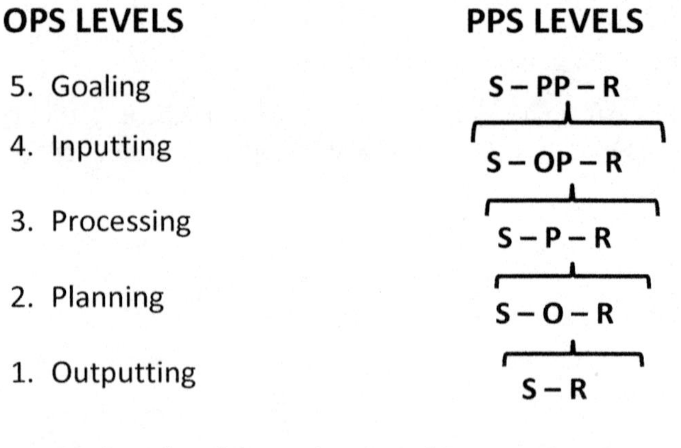

Figure 23. Levels of Organizational Processing Systems

OPS 2. Transferring to Organizational Functions

The second step is to transfer the phenomenal processes into organizational functions. Just as PPS identifies its phenomenal operations, so does OPS identify its organizational functions derived from **Marketplace Capital Development (MCD) Systems** (see Figure 24).

OPS PROCESSES		MCD FUNCTIONS
Goaling	5	Policy (Mission)
Inputting	4	Executive (Architecture)
Processing	3	Management (Systems)
Planning	2	Supervision (Objectives)
Outputting	1	Delivery (Tasks)

Figure 24. OPS Processes and MCD Functions

OPS 3. Transferring to Organizational Components

In a similar manner, the OPS processes are transferred into **Organizational Capital Development (OCD) Components** (see Figure 25).

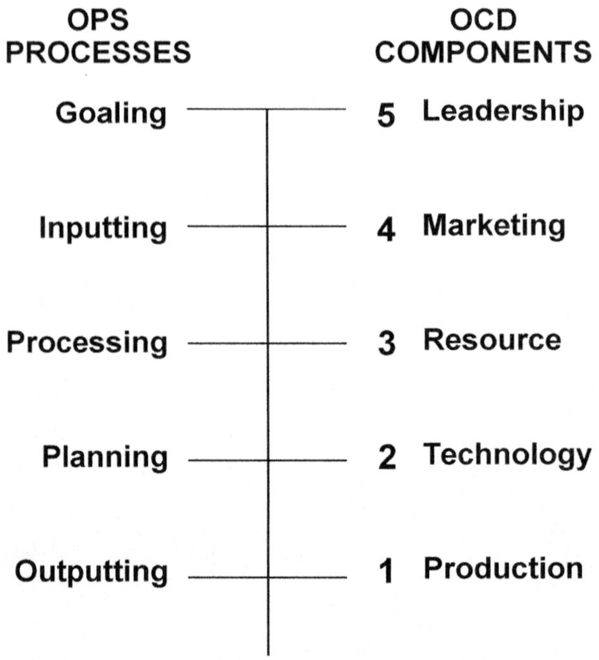

Figure 25. OPS Processes and OCD Components

OPS 4. Relating the Phenomenal Paradigms

The fourth step is to relate the phenomenal paradigms. As may be viewed in Figure 26, the OPS components are aligned to discharge the OPS functions.

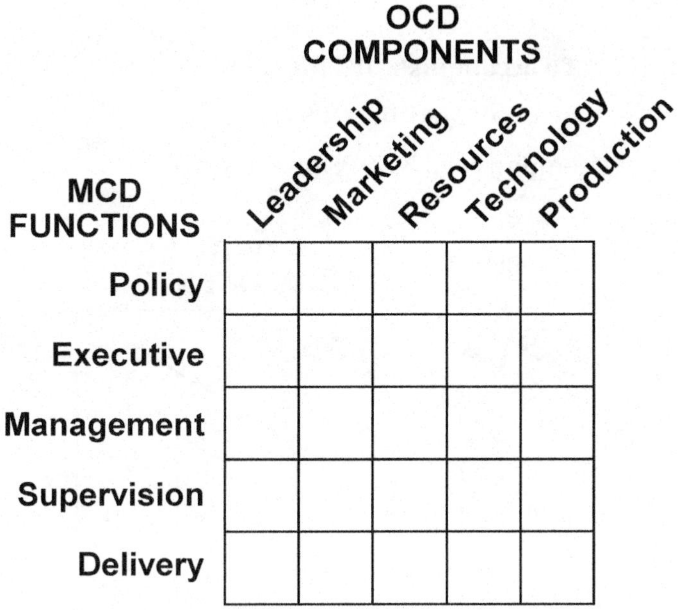

Figure 26. OCD Components and MCD Functions

OPS 5. Empowering the Phenomenal Paradigms

The fifth step is to empower the related paradigms with the generating engine. As may be viewed in Figure 27, the OCD components are empowered by the OPS processes to discharge the MCD functions. Integrated, these dimensions define the OCD mission:

To accomplish OPS functions

by implementing OPS components

empowered by OPS processes.

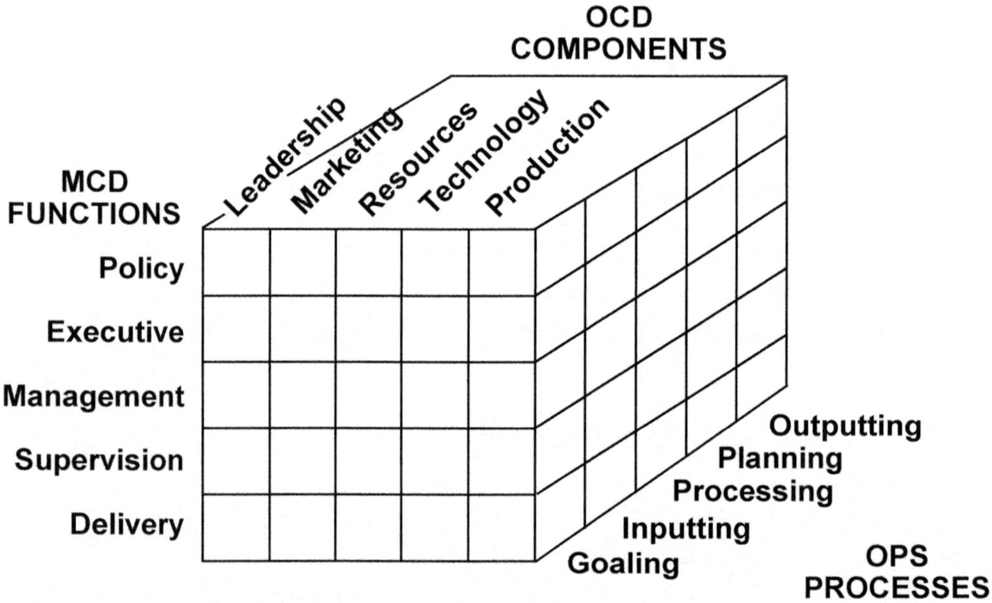

Figure 27. The Comprehensive OPS Model*

*The Organizational Firing Order is available for viewing in animation on Pod IV of **mcleanproject.com.**

10

The Phenomenal Processing Paradigm

The **Phenomenal Processing Paradigm** or **PPP** applies to all human phenomena. As may be viewed in Figure 28, we may derive Human Capital Development (HCD) models from Organizational Capital Development (OCD) models simply by "rotating" the dimensions deductively or counter-clockwise.

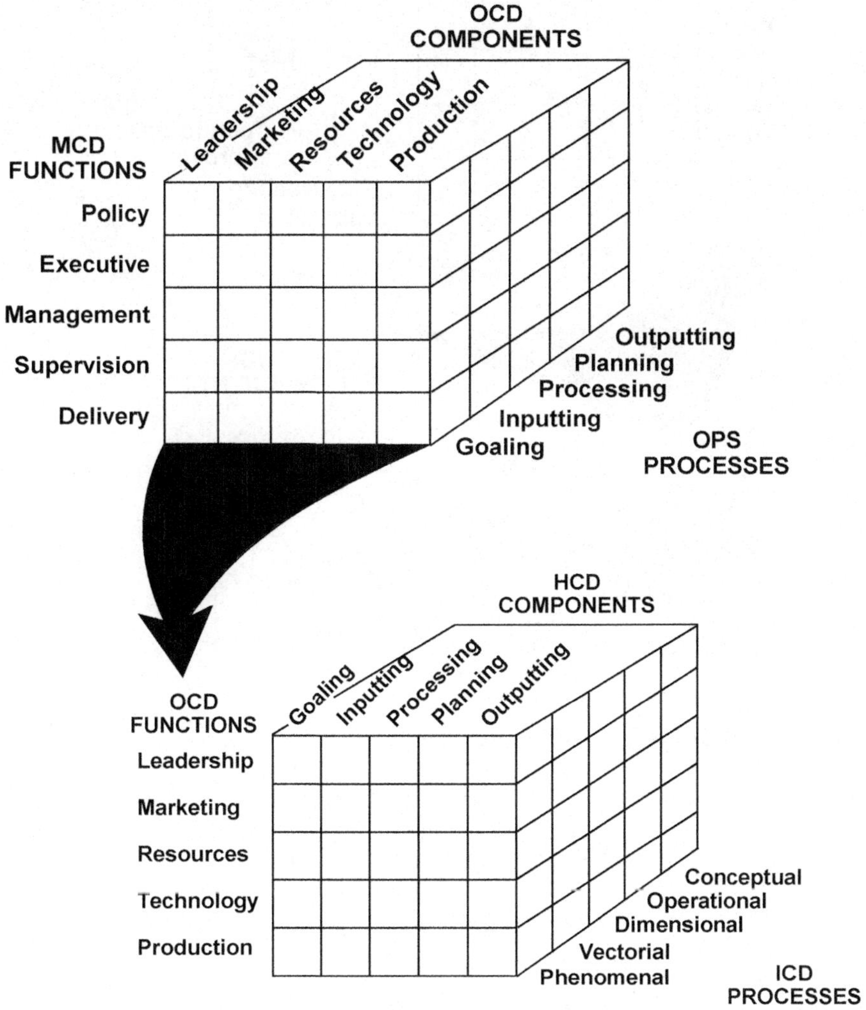

Figure 28. OCD Generates HCD Models

In a similar manner, we may generate **Marketplace Capital Development** or **MCD** by "rotating" inductively or clockwise from OCD to **Community Capital Development** or **CCD** (see Figure 29).

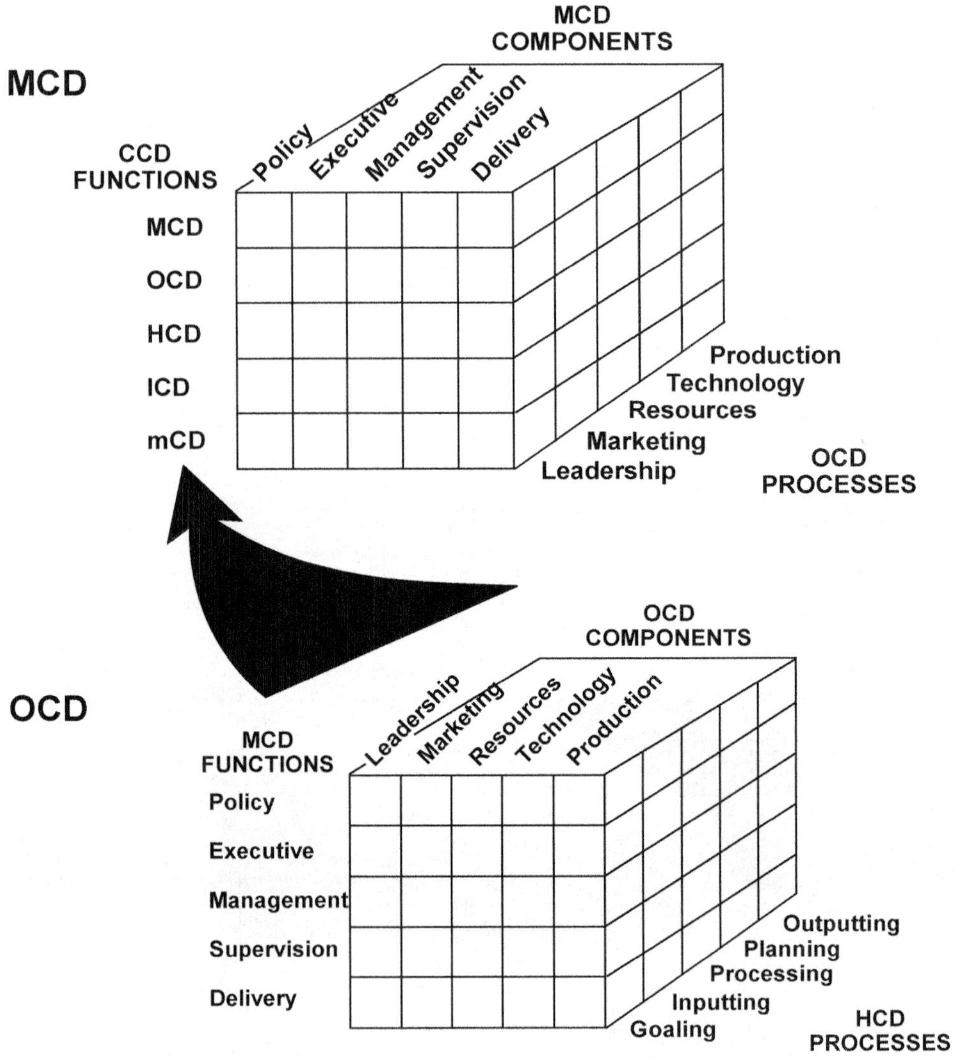

Figure 29. OCD Generates MCD Models

Deductive Generativity

We may view deductive generativity in operation in Figure 30 below. Again, the OCD dimensions are "rotated" deductively and counter-clockwise to serve new operations in lower-order "nested" systems. As may be viewed, each **New Capital Development** or **NCD** model derives its "rotated" dimensions from the higher-order model.

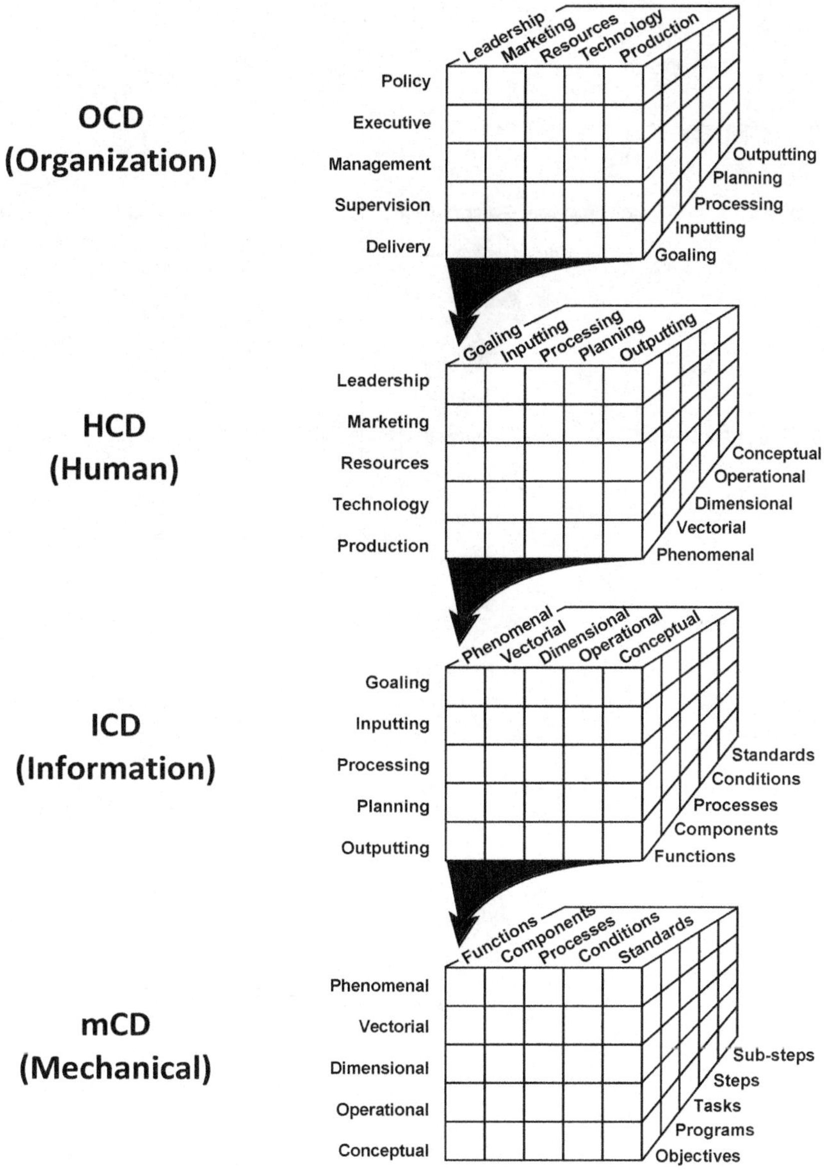

Figure 30. New Capital Development Systems

Inductive Generativity

In a similar manner, the MCD dimensions are "rotated" inductively and clockwise to serve new operations in the higher-order "nesting" systems (see Figure 31). As may be viewed, each MCD Capital Development model derives "rotated" dimensions from the lower-order models.

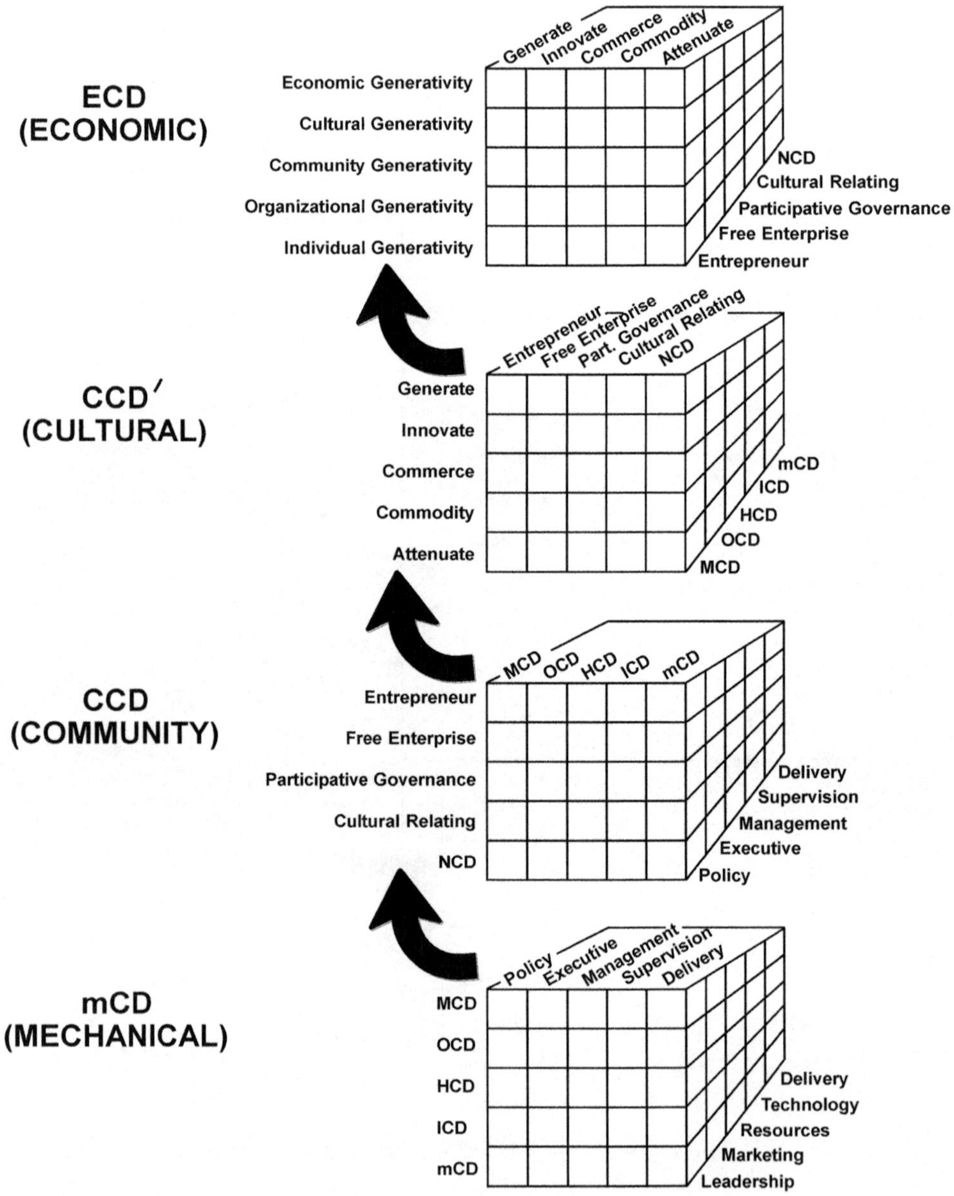

Figure 31. Socioeconomic Growth Systems

"PHEN-GEN" Systems

Functionally, the ODNA systems have transformed into **"O-GEN"** or **Phenomenal Generativity Systems.** The **"PHEN-GEN"** systems may generate any capital development systems. In these illustrations, we have generated the following models and their functions:

- The **ECD** model for socioeconomic development is the most powerful capital development model of our phenomenal systems.
- The **ECD** model "nests" and "encodes" all other capital development models:
    - CCD′—Cultural
    - CCD—Community
    - MCD—Marketplace
    - OCD—Organization
    - HCD—Human
    - ICD—Information
    - mCD—Mechanical
- All models may be rotated to accomplish any capital development functions.

We can now generate the capital development missions of all **"rotated"** models:

ECD (Economic)	Economic generativity functions are accomplished by generative-driven economic components empowered by entrepreneurial-driven cultural processes.
CCD′ (Cultural)	Cultural generativity functions are achieved by entrepreneurial-driven components empowered by **MCD**-driven positioning community processes.
CCD (Community)	Entrepreneurial functions are achieved by market-driven community components empowered by policy-driven organizational alignment processes.
MCD (Marketplace)	Market capital development functions are achieved by policy-driven organizational components empowered by leadership-driven processes.
OCD (Organization)	Policy-driven marketplace functions are achieved by leadership-driven organizational components empowered by goaling-driven human processes.
HCD (Human)	Leadership-driven organizational functions are achieved by goaling-driven human components empowered by phenomenally-driven information modeling processes.

ICD (Information)	Goaling-driven human processing functions are achieved by phenomenally-imaging information components empowered by functions-driven mechanical tooling processes.
mCD (Mechanical)	Phenomenally-imaged information functions are achieved by functions-driven mechanical tooling processes empowered by objectives-driven sub-mechanical processes.

Summary and Transition

In summary, we may employ the "PHEN-GEN" systems to generate any capital development models. "PHEN-GEN" emphasizes **"rotating"** or **"impregnating"** robust DNA to generate new forms of capital development models.

In transition, "PHEN-GEN" expedites the transfer of prepotent **"dimensional genes"** for the generation of any new forms of capital development:

PHEN-GEN → *n*! (nest! encode! rotate!)

"PHEN-GEN" can generate the architecture for any place in our multiple universes—internal or external.

Part V

Community Generativity

11

The Generative Community

Dick Sprinthall was the initiator of "The Community Project" and therefore had a much larger perspective than all of us. His mission was "Saving Springfield." In his words, "It is already one hour to midnight."

In this regard, Sprinthall had several unique perspectives on the crises:

- As professor of psychology at American International College, he understood the "island" around which the underclass black and brown communities had congregated.
- As mentor of Andy Griffin, a graduate of the college and the recognized leader of the black community, he understood their sense of hopelessness.
- As consultant to the CEOs of the two biggest corporations in town, Jim Martin of Mass Mutual Life Insurance and Ben Jones of Monarch Life Insurance, he understood corporate fear of being surrounded by the underclass minority communities.

Sprinthall's design was simple. Bring in the most powerful human and community resource developers, Carkhuff and Berenson, and unite them with the most powerful black community organizer, Griffin. The results that he foresaw would fit the "win–win paradigm":

- The college would "win" with the reflected benefits, including financial support that would attend their contributions to the resolution of community crises.
- The members of the black community would "win" by elevating their educational and vocational qualifications and acquiring jobs and earnings.
- The corporations would "win" by making peace with the underclass communities that surrounded them.

All of these "wins" would come to pass with the resolution of community crises. But it was not to be without other crises that upped the ante for the participants involved!

What Sprinthall had not fully recognized was the precarious nature of the crises:

- A community rebellion of the underclass was coalescing, led by young black militants who were prepared to die in the streets.
- An investor rebellion was silently occurring that called for the corporations to move their elaborate enterprises into the suburbs away from Springfield.

The net effect of civil disturbances would be to lose both the diminishing sources of wealth for the community and the disappearing sources of hope for the people. As industries fled to more cost-beneficial climes, the people who would suffer most were the disenfranchised underclass populations with increasingly higher rates of unemployment, welfare, criminality, and other indices of dysfunctionality. There were already proposals for "ghettoizing" the underclass communities with "walls and cops." The police already had the rebel leaders "in the crosshairs." Springfield was indeed in a precarious condition.

It was already ten minutes to midnight!

In other words, the conditions for both the corporations and the people who surrounded them were deteriorating and exacerbating. Consequently, the requirements for success were elevated. "The grow–grow paradigm" was born.

"The grow–grow paradigm" began with an understanding of the fruitlessness of "win–win" when neither of the parties had anything to offer. "Grow–grow" was generated as follows:

All parties must grow before either party can relate.

Like people in conflict, all parties must elevate before any parties can relate.

Accordingly, we defined our strategic objectives as follows:

1. To empower the corporations to grow
2. To empower the community to grow
3. To relate each to the other

Concerning the corporate objectives, we invited them to participate in community development seminars in our graduate program where we explored and initiated projects for all aspects of community development. We empowered the corporations to make tangible internal as well as measurable community innovations. Perhaps most dramatic at this time, we introduced their corporate leaders to the use of computers in their actuarial-based insurance industries. In so doing, we contributed dramatically to the future growth of their industries.

Concerning the community objectives, we empowered waves of candidates with educational credentials and job skills for both the private and the public sectors. In this context, the program graduates "lobbied" the local corporations for entry-level positions with the promise that their future performances would be "interchangeable" with those of the current corporate employees. This proved to be the case. We contributed dramatically to the future growth of the community.

Most important, we brought representatives together periodically in "Real People's Congresses" where all parties involved shared their problems and co-processed to generate solutions. Together, we designed and developed the next sources of industrial and community development:

- Information technology-driven
- Human technology-supported

To be sure, "The Springfield Miracle" was born and transported across the state to Boston and other communities, which, supported by the IT development in and around Route 128, soon evolved into "The Massachusetts Miracle."

Perhaps most important of all was the birth and growth of the generative processing systems that were dedicated to thinking outside the box. These systems generate solutions to human problems that are not currently available.

For example, the most problematic of all areas of human endeavor is that of developing human potential. Many of us have been conditioned to believe in the inherent superiority and inferiority of different populations. Indeed, it is our own conditioning that has led us to believe that the underclass is genetically determined.

Generativity is born in the community along with the infants of all sorts of homes and families. No child is excluded from the potential for generativity by their birth-families, schools, communities, cultures, or nations. The critical issue here is this: what do the families, schools, communities, cultures, and nations do to stimulate, facilitate, and actualize the potentially unlimited brainpower with which each child is born?

Figure 32 illustrates the basic community through which an individual takes a journey:

- **The homes and neighborhoods** prepare the infants with linear conditioning and mechanical responses.
- **The schools and training** empower the children with discriminative learning skills and introduction to information responses.
- **The colleges and technology centers** empower the students in generative individual processing with an introduction to human processing systems.
- **The government and service agencies** empower the citizens in generative organizational processing with an introduction to organizational processing systems.
- **Business and industry** empower their employees in generative marketplace processing with an introduction to marketplace and environmental processing systems.

As may be noted, each of the community components may be related to the other components. This is the key to effectiveness in the "Generative Community."

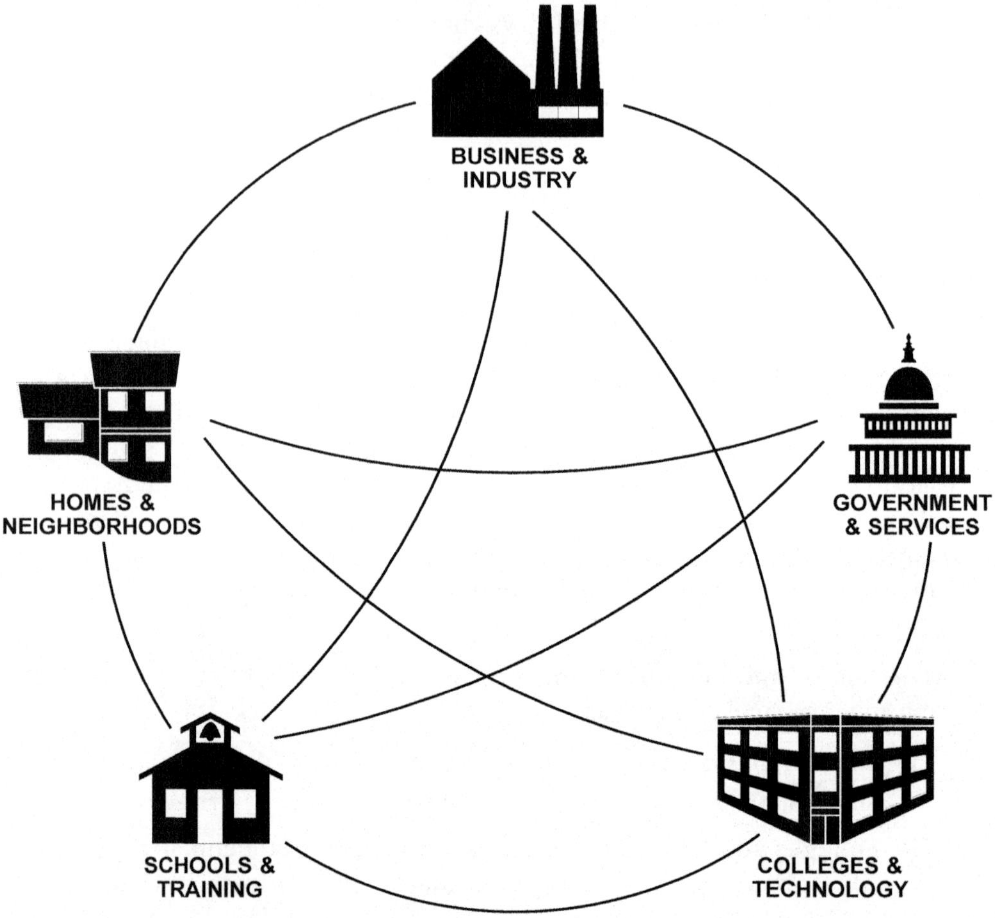

Figure 32. The Generative Community

12

Generative Homes

History is replete with studies of the effects of neglect on child development. Scientists, psychologists, and nutritionists have all concluded that neglect is a powerful "depressor variable." When neglect intervenes in a child's life, the effects are disastrous: physically, emotionally, and intellectually. Clearly, the environment in the form of parents and their surrogates did not stimulate and support these children. Their parents did not know how to do so because no one had ever taught them how to do so. They could not teach their children because no one had ever taught them. They could not support their children's learning because no one had ever supported their own learning.

The neglected children, many of them superior in intellectual resources, were destined to live with minds and bodies enslaved by intellectual limitations just as surely as the children's ancestors had been enslaved by chains. Many of these children were to subsist below poverty levels on their poor farms. Or they were to move to the city and join the ranks of the unemployed. Or they were to become less than full persons because of financial dependence on the welfare system. Or they were to opt for a criminal career ladder and end up in prison or the morgue.

All because we did not help them to actualize their superior potential. And no amount of guidance will serve to guide them down the path to education, no amount of rehabilitation counseling will help them to actualize their potential, no amount of psychotherapy will help them to emerge as full human beings, no amount of correctional counseling will give them back their freedom. Because we are not ready for our own fulfillment, our own actualization, our own emergence, our own freedom, we will not give these children theirs.

Some people rise above their bonds to study the effects of a human environment on little human beings. This brings us to the studies of children in Milwaukee by Rick Heber and associates. In the 1970s, they studied the relationship between poverty and mental functioning. It divided the children of intellectually challenged welfare mothers into two groups. In the experimental group, the researchers carefully selected teachers and trained them to work with the infants of the mothers on welfare. The teachers, in turn, trained the mothers of infants to respond sensitively to the children and to stimulate and reinforce their learning activities.

In other words, all parties involved worked together to maximize the impact of the environment at the point where the child was going through the greatest growth. And they did this through skills training. Skills for the teachers. Skills for the parents. Skills for the children.

What do you suppose the difference in IQ was between the experimental and control infants in the welfare study after three-and-one-half years?

Ten points? Twenty points? Thirty points?

The differences between the children who were nourished intellectually and those who were not was 33 points.

Think of it! The study focused on children who, in the absence of continuous responsiveness and stimulation, are destined to live out their lives as city slum dwellers at best—and at worst, may have no lives at all!

The children in the welfare study, in the presence of continuous responsiveness and stimulation, may exercise their superior abilities to rise above their places of origin.

The difference?

The enslaved group was abandoned to the "benign neglect" of the playpen and the TV set. The free group was saved by committed and disciplined people who put the human being back into the learning equation. But most of all, the free group was saved by skilled people who taught skills to people who, in turn, taught skills to other people.

In the instance of neglect, the unattended children may drop by more than 30 IQ points over a period of several years. In the instance of facilitation, the attended children may gain more than 30 points over a period of several years.

The difference between facilitation and neglect is the difference between life and death itself. And it is the use—or the absence—of skills that makes all the difference!

Developing Human Potential

The learning journey begins in infancy. It extends through early childhood in the homes and neighborhoods, in the family processes, and in extended family relations. The homes are the most simple unit of basic community development (see Figure 33). There the effective parents or surrogates are dedicated to empowering their children to learn how things work. We label these functions "mechanical" and define them operationally as **Mechanical Capital Development** or **mCD.**

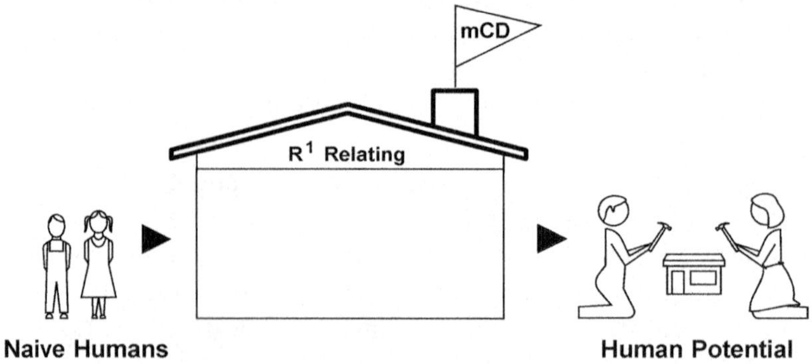

Figure 33. Building the Operations of the Homes

To this function, the parents dedicate their **R^1 – Relating Skills:** they relate to each other and their children; they relate the children to the environment; the children learn how things work.

In terms of the home operations, then, the untutored children are transformed into tutored children with "mechanical learning." In other words, the tutored children can both discriminate and communicate an understanding of how the world around them works.

The "mechanical learning curriculum" emphasizes the "basic interrogatives":

- **Who** is involved?
- **What** are they doing?
- **Why** are they doing it?
- **Where and when** are they doing it?
- **How** are they doing it?
- **How well** are they doing it?

We label these interrogatives "5W2H" to reflect our understanding of how things work.

The parents employ two sets of critical relating skills in teaching their children:

- Relating or responding empathically to the child's frame of reference
- Relating or initiating to answer the basic interrogatives defining the world around them

In short, the parents as teachers transform their children from naïve children into mechanically skilled (mCD) learners. By so doing, the parents elevate human beings into powerful "human potential." Moreover, like the children in the welfare study, the parents transform their children into persons of unlimited potential as they prepare them for the "voyage of discovery" ahead. Until we, as parents, learn to behave by relating to our children and their worlds, we are but part of the same slave mentality that enslaves the masters as well as the slaves.

13

Generative Schools

All educators are familiar with the "achievement deficit." With each year of schooling, the black children fall further and further behind on the "achievement curve."

Well, some black parents worried long and hard about it. And they were determined to do something. There were black leaders who taught them that if black people get involved, their kids will learn fast. Back in the 1990s, the people of Chester formed the first charter school in Pennsylvania: The Chester Village Charter School. They formed a black school board. They appointed a black headmaster. They increased the ranks of black teachers to teach their 100% black students. And they fell further behind on the "achievement curve!"

Falling Further Behind

Not only did they fall behind in achievement but also in attendance, deportment, and all other indices of functionality. Daily, when arguments broke out at the lower class levels, the upper class students emptied their classes and brawled in the halls. Teachers, themselves, were not safe from assaults and flying missiles like erasers and chalk. Many of us know the scene because we saw it on *The Wire.*

Outside the school, the reputation was as a "feeder school" to the violent public high school. The "Boyle Street Gang," graduates of the Charter School, served as "Murder, Inc." for the City of Chester: just leave $100 in the right place and you could have anyone assassinated!

To be sure, 85% of the boys over 16 years of age were in the criminal justice system. Nearly two-thirds of the girls over 16 were pregnant and on welfare. Things were out of control!

The leader and the teachers were in a depression because the kids were not learning the way they should. They felt that they had done everything that they could do naturally. And they cared! Above all else, they cared! Yet things got "worser" as they say.

Most of all, the kids were in a state of agitated depression. They came in innocent and hopeful. They opened to an experience that they found harmful. They left street-wise and fearful.

Some dropped out. Others "fell out." Most just "passed out"—physically, emotionally, and intellectually. Those who "passed on" found that the Village Charter School had been a training ground for the Chester Upland Public High School: it prepared the students for the retarding experiences that lay ahead.

Getting Ahead

501 out of 501: The charter school got its rankings in the State of Pennsylvania. It was dead last in achievement. Things could not get any worse. It had become the highest priority for closing in the state—"The failed charter school."

Now the school faced the feared standards of "No Child Left Behind," the controversial federal legislation. The school's goal was to make "AYP," or "Adequate Yearly Progress," which none of the schools in the district had made before.

Yet the leaders never gave up! They internalized their own inadequacies. They had no finances for consulting and training, so they could not even find prescriptions for the limitations that they diagnosed. Nevertheless, they sought help.

The help came in the form of "The Possibilities Schools," which introduced a new curriculum—literally "The New 3Rs."

"The New 3Rs" were dedicated to teaching the learners to think. In terms of the conventional curriculum of "The Old 3Rs," the learners were taught to co-process the content with the teachers by employing "The New 3Rs":

- **Relating** to exchange images
- **Representing** images in operational forms
- **Reasoning** with the new images

Both kids and teachers responded with enthusiasm; all had found a new and "awesome" way of learning. They responded accordingly with awesome effort.

The most important thing that "The Possibilities Schools" program did was to install one of those headmasters who never gave up believing in the kids. His name was John Linder and he was, himself, a "homeboy" from Chester who never gave up hope. He was empowered in "The New 3Rs" and he took an incredible journey into teaching effectiveness. "The Possibilities Schools" now had the model they needed for everyone—teachers and students alike—to emulate.

The second most important thing that "The Possibilities Schools" program did was to install Dale Kelly as the assistant headmaster. Ms. Kelly acted as a "lead-teacher" who managed the entire school as if it were her class. She loved every child as her own and empowered every teacher as agents of the children. Indeed, "The Possibilities Schools" now had the agent that they needed to learn from.

Empowered with both model and agent for "The New 3Rs," the students of the Village Charter School went on to break all records for performance. To be sure, it became the only school in the district to accomplish the coveted AYP—above 10% overall.

From 501 to AYP! Here is the rap the students put together and recited:

"The Relating Rap"
by the Students of Chester Village Charter School

I. Village Charter is The Relating School.
It teaches us The Relating Rule—
Relate to people and relate to knowledge,
Relate to skills that get us into college.

II. Above all else, we relate to read.
To decode our reading, we break down the deed.
Our guiding rule is "Basic Interrogatives."
That breaks down into "Viable Perogatives."

III. "Who is doing what to whom?"
That's the basic question we have to assume.
"How and why are they doing their thing?"
"Where and when?" answers everything.

IV. We relate the same in order to write,
Who, what, when, where, how and why?
"5WH" is our formula.
It makes performers out of us all.

V. When we began we were the worst,
Now we know we are the first!
"We're Number One," that's just deserve,
We're getting ready to jump the curve.

VI. Village Charter is The Relating School.
It teaches us The Relating Rule.
We relate to people and relate to knowledge.
We relate to skills that get us into college.

Developing Human Resources

The learning journey continues through childhood and adolescence in the schools and training centers. The schools are the next level of complexity in developmental and cumulative building (see Figure 34). There the effective teachers and trainers are dedicated to empowering the learners in representing images of information at all levels. We label these functions "informational" and define them operationally as **Information Capital Development** or **ICD.**

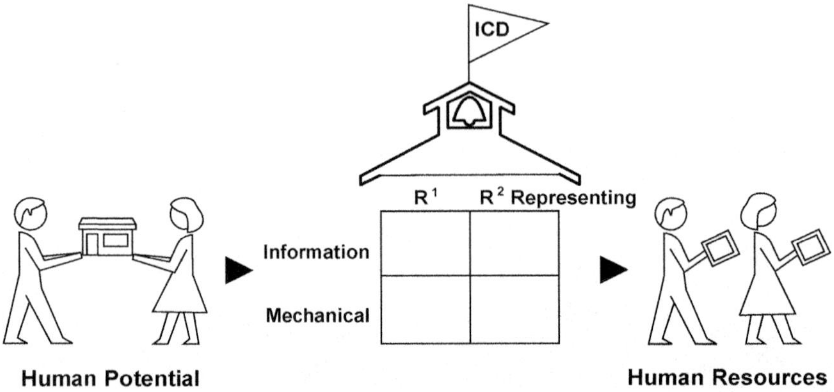

Figure 34. Building School Operations

Cumulatively, then, the teachers dedicate their **R^2 – Representing Skills** to the ICD functions: they relate and represent images of the experiences of the learners; they relate and represent images of the phenomena in their worlds. For example, the learners have already represented images in terms of the basic interrogatives in simple sentences. At higher levels, they may represent images in operational systems incorporating those images.

In terms of the school operations, then, the learners with human potential are transformed into human resources with "Information Representation Skills." In other words, the human resources can process images into elevated levels of information. That is precisely what makes human resources valuable: the ability to discriminate different levels of information.

The "ICD Learning Curriculum" emphasizes the images of systems:

- **Contexts emphasizing conditions and requirements**
- **Inputs emphasizing parts or participants**
- **Processes emphasizing procedures and methods**
- **Outputs emphasizing performance and products**
- **Feedback emphasizing measurements**

We call these images "Operational Systems" because they define programmatic ways of accomplishing any goals.

The teachers, then, employ two sets of critical skills in teaching the students:

- Relating to the learners' experiential images
- Representing the phenomenal images of the world around them

In short, the teachers or trainers transform the learners from human potential into human resources by ICD processes. By so doing, the teachers elevate human potential into potent human resources.

Again, like the children in the Milwaukee welfare study, the students now have infinite potential as they engage in the early phases of their "voyage of discovery."

14

Generative Colleges

Of all the "Preconditions of Civil Disorder" defined by the Kerner Commission as the essential ingredients for civil disturbance, the most fundamental is the racial attitude and behavior of white Americans toward black Americans. Racial prejudice has shaped our history decisively; it now threatens to affect our future. The key to unlocking the doors of community crises is empowerment in generative processing.

Civil Disorder

Among the "Preconditions of Civil Disorder", the following are most critical:

- Pervasive discrimination and segregation in employment, education, and housing, resulting in continuing exclusion of great numbers of blacks from "The American Experience"
- Social, political, and economic conditions of inner-city communities constituting a "clear pattern of severe disadvantage for blacks compared with whites"

Indeed, the frustrations of powerlessness had led some blacks to the conviction that there was no effective alternative to violence as a means of achieving redress of grievances and of "moving the system." These frustrations were reflected in alienation and hostility toward the institutions of law and government as well as the private sector of the white society that controlled them. To be sure, it is the great fear of "home-grown terrorists" that will drive the intelligent majority community to once-again invite the minority community to participate in the "American Experience."

At American International College, leaders of both the black and white communities were recruited to be empowered in processing interdependently the ongoing crises of their time. In other words, the empowered participants processed in "real-time" to generate solutions.

The American Experience

Clearly, the solution to exclusion is inclusion: full participation in the mission of the American Experience:

- Cultural relating
- Participative governance
- Free enterprise

These became the goals of the first Center for Human Resource Development (HRD). The programs were based on the "Core of Generativity":

- Relating to images
- Representing images
- Reasoning with images

We began with the human relations skills that enable us to relate not only between groups but within our own groups. Over a month-long program, we divided blacks and whites into groups. Within each group, we empowered the members to relate effectively to each other with interpersonal communication skills. Next, we brought the groups together to relate between groups. The cultural relating was immediately successful, and the participants were ecstatic. Cultural relating is the cornerstone of all human endeavor. It led us directly to our next steps.

Since the majority community addressed the minority community with "benign neglect," our "Real People's Congress" committed immediately to defining and implementing a "shadow government." Our community leader, Andrew Griffin, became our "mayor pro tem." His mission was to engage all members of the community with one basic question:

"What would you like to do with the rest of your life?"

They answered universally:

"We would like to be everywhere that impacts us."

An incredibly profound answer! We had our assignments. We were off-and-running to meet the needs of our citizens. Participative governance is the core ingredient of the American Experience. It led us directly to our next steps.

Basically, the responses of thousands of citizens directed us to initiate entrepreneurially in both the public and private sectors. We developed our programs in the terms that the people conceived—"from womb-to-tomb!"

- Pre-parenting skills
- Early childhood preparation
- Primary and secondary education
- Private and public sector employment
- Participation in governance
- Participation in politics
- Participation in integrated housing
- Participation in recreation and fitness
- Participation in counseling and guidance
- Participation in bereavement and other benefits

Obviously, in order to accomplish these objectives, we had to define and implement new roles for ourselves. Not only did we need to define new areas of thinking and relating, we also had to implement new areas of lobbying and publicizing!

The most powerful outcomes were professional:

- In terms of cultural relating, all graduates attended college, and more than 50% were graduated (this from a population with an average seventh-grade education).
- In terms of participative governance, all graduates participated in city government with many assuming leadership roles (Ray Jordan became a Massachusetts state senator and chaired the Committee on Ways and Means).
- In terms of free enterprise, the graduates out-earned the majority community (three of the graduates became millionaires).

Perhaps the most dramatic outcomes were personal as well as beneficial to the community:

- None of the men ever committed another crime (all had criminal records before the HRD program).
- None of the women ever went on welfare again (all had been on welfare before the HRD program).

The formula was straightforward:

> Engage the disenfranchised in the American Experience and they will respond as all other Americans have:

- **Relating for Peace**
- **Enlightening for Participation**
- **Enterprising for Prosperity**

The HRD graduates will never stop growing! Generativity is a spiral that never ends!

Generating Human Capital

For the individual, the learning journey culminates in colleges and technology centers. The colleges are the next level of complexity in developmental and cumulative building (see Figure 35). There the effective professors and laboratory leaders are dedicated to empowering the post-secondary students in reasoning to generate new and improved images of information. We label these functions "human generativity" and define them operationally as **Human Capital Development** or **HCD.**

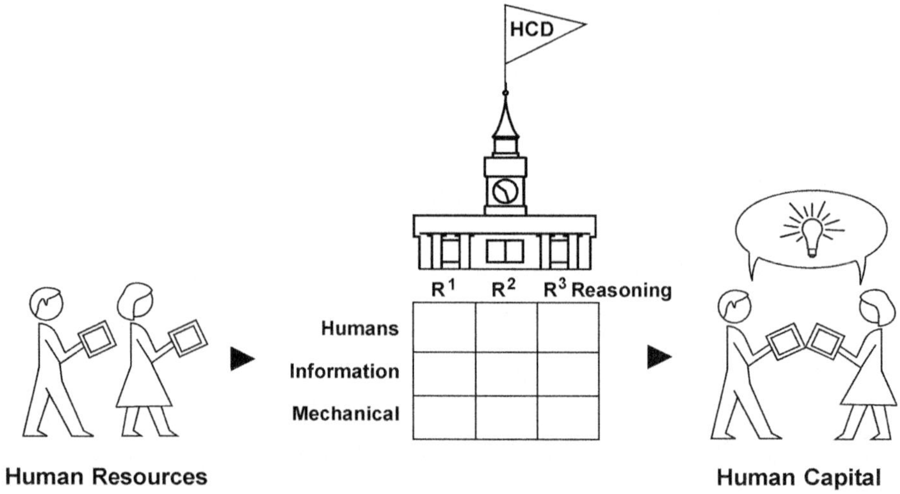

Figure 35. Building College Operations

Cumulatively, then, the professors dedicated their **R³ – Reasoning Skills** to the HCD functions. After relating and representing images, they explore, understand, and act on new images of phenomena in their worlds. For example, the students may have represented images in operational systems. In reasoning, they may explore by expanding to multidimensional images before understanding by narrowing to preferred images for achieving our values and the condition's requirements.

This is precisely why we label R³ – Reasoning a generative process: we create new and more productive images by expanding and then narrowing new images before acting on them. When we think generatively, we are thinking outside the box or beyond the high beams.

In terms of college and technology operations, then, human resources are transformed into human capital with "human generativity skills." In point of fact, the generative process is actually a "synergistic process": both human and information capital grow synergistically or in relation to each other's growth. This is what makes "humans" into "capital" (e.g., meaning "what is most important").

Model-building in the "HCD Generating Curriculum" emphasizes the relationships of images of schematic relationships between systems:

- **Components** emphasizing parts
- **Functions** emphasizing outputs
- **Processes** emphasizing procedures
- **Conditions** emphasizing requirements
- **Standards** emphasizing measurements

We call these images "dimensional systems" because they define the systematic ways of accomplishing any missions.

The professors, then, employ three sets of critical skills in empowering the students to think generatively:

- **Relating** to the learners' experiential images
- **Representing** the images of the phenomena they are addressing
- **Reasoning** to explore, understand, and act on the new and more productive images

In short, the professors transform the students from human resources into human capital by HCD processes. In so doing, the professors elevate the human resources into human capital with "self-generating functional autonomy."

Most powerfully, the **"Generativity Curriculum"** was exported across the country:

- **Captain Charles Montgomery, a graduate of the "First Correctional Counseling Institute" conducted in Springfield, established the first "Center for Correctional Counseling" for the Federal Bureau of Prisons with outstanding results.**
- **Dr. Terry Bergeson, Superintendent, Washington schools, along with her mentor, Dr. Griffin, generated the most outstanding state school system in the country with SAT performance, student retention, and other indices of student learning.**
- **Dr. Griffin transferred the basic curricula to the Boston vicinity, thus contributing to the generation of "The Massachusetts Miracle" of the 1988 political campaign.**

Once again, like the children in the Milwaukee welfare study, the students now have infinite potential for generativity in their "voyage of discovery."

15

Generative Government

Back in the late 1960s, the HRD Center in Springfield developed a graduate program in Human Resource Development—Community Resource Development (HRD–CRD). More informally, this program was known as "The Shadow Community" installed because of the "benign neglect" that the majority community and its governance reflected on their marginal black and brown citizens.

The Shadow Community

The mission of "The Shadow Community" was to bring together representatives of all organizations affecting and affected by HRD and CRD. Accordingly, the HRD Center recruited heavily from leaders of the private sector as well as leaders from the public sector and the community-at-large. In their regular sessions, they processed interdependently the critical issues of their time: social, economic, employment, educational, and other areas.

For each problem area, the CRD group generated solutions that they then acted on. For example, to improve deteriorating socioeconomic conditions in the inner-city, the CRD group contributed to the architecture of "Bay State West," a revitalization of business development in the core of the inner-city. To improve employment and economics, the CRD group lobbied both the private and public sectors for entry-level positions for which the HRD program was responsible for training and supporting. Educationally, the CRD group initiated Educational Achievement Programs for those left behind by conventional public education (e.g., new careers programs for the old career drop-outs). Along these lines, "The Concentrated Employment Program" was converted to a college-degree preparation program in which "90-day wonders" received training in core living and learning skills and specific working skills of their individual choices.

Programs in many other areas such as housing and welfare and criminal justice all followed with astounding results. A 20-year follow-up indicated the following enduring highlights of performance:

- All graduates were gainfully employed, many at executive, management, and supervisory levels, including one who ultimately sat on the Board of Directors of a Fortune 500 firm.
- All graduates went on to complete at least one year of college at many different post-secondary institutions, including one graduate with a doctorate from Harvard.

Moreover, the college in which the graduate program was housed received due recognition. While all colleges in New England experienced significant declines in applications for admissions in the late 1960s and early 1970s, during the years the HRD Center was in operation, American International College experienced an increase of 15% to 20% in applications. According to J. Walter Reardon, vice president, Mass Mutual, and a graduate student in the program, "This was a million-dollar public relations miracle. It was as if parents and their children were saying, 'At least AIC is dealing with the relevant problems of our time.'"

The demonstrations were important: marginalized people really can participate and perform in the American Experience.

The implications are profound: what is the majority doing preventing the minority peoples from participating in the American Experience?

Perhaps the most important demonstration of the CRD group was the introduction to computerized contributions to be performed. In these group meetings, Dr. Bernard G. Berenson, a pathfinder in computer applications, introduced the groups to the use of computers. Based as the major industries were upon actuarial data, the computerization came to spur the revitalization of Springfield industry.

In summary, the major facets of the American Experience were processed in the CRD group in the HRD Center:

- Cultural relating in relations within, between, and among the private, public, and community sectors
- Participative governance in responding and then initiating proactively to the crises within, between, and among the private, public, and community sectors
- Entrepreneurial enterprise in generating, risk-taking, and managing new initiatives leading to new enterprises within, between, and among the private, public, and community sectors

The participants in the CRD group had indeed designed and implemented the Shadow Community. In so doing, they, themselves, were transformed from prime human capital to "enlightened citizens."

Generating Enlightened Citizens

For the individual, the learning journey is accelerated in government and organizations of all kinds. The governmental agencies are the next level of complexity in developmental and cumulative building. The governance agencies introduce the organization to the community (see Figure 36). The productive leaders and agency directors are dedicated to empowering "generative thinkers" to reason generatively in an organizational context. We label these functions "organizational generativity" and define them operationally as **Prime Human Capital** or **Enlightened Citizens.**

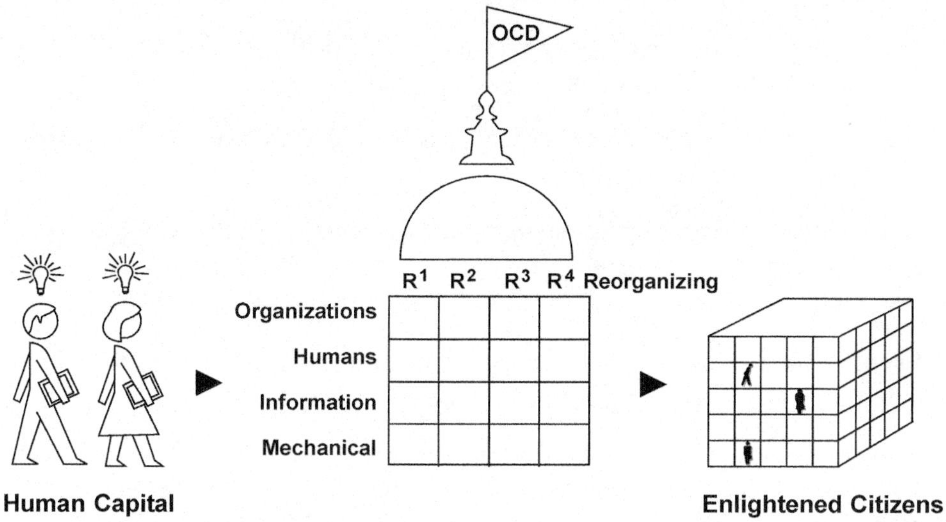

Figure 36. Building Governance Operations

Cumulatively, then, the leaders and directors dedicate their **R⁴ – Reorganizing Skills** to Prime HCD functions. In this context, the generative processors dedicate their R³ – Reasoning processes to R⁴ – Reorganizing functions. For example, the generative processors may generate new and preferred representations of images for organizational processes: Goaling, Inputting, Processing, Planning, Outputting.

This is precisely why we label R⁴ – Reorganizing an organizationally generative process: we generate new and more productive representations of organizational images for the continuous alignment of organizational resources. When we think generatively in the organizational context, we are thinking outside the "organizational box": we view our past and current requirements and capacities in order to formulate our future objectives.

In terms of governmental and organizational operations, then, human capital are transformed into prime human capital or enlightened citizens with "organizational generativity skills." In practice, the generative processors engage in "synergistic processes" with the organization. Processing generatively with the organization, they become "one" with it. This "oneness" of processing is what transforms human capital into **Organizational Capital.**

Important demonstrations of government generativity that began in Springfield extended to the federal sector:

- **The creation of the first Indefinite Quality Contracting mechanism in the federal sector for the Office of Personnel Management.**
- **The creation of the first profit center, Training and Management Assistance Program, in the federal sector.**
- **The architecture for the first Instructional Systems Design to empower all agencies in the federal sector.**

To sum, graduates of the HRD program were central contributors to the introduction of HCD, ICD, and OCD programs into the federal sector.

16

Generative Business

After all is said and done, it is not the energy crisis that has put America on the edge of bankruptcy—economic, social, and moral. It is the "generativity crisis"!

This crisis in generative thinking spreads virus-like, just below the surface, blurring our abilities to observe and diagnose. Osmotically seeping through our permeable membranes, this virus penetrated the very thought processes of our brains.

The consumption virus brought with it a continuous and incremental reinforcement schedule of benefits—of "quick hit" innovations and successes. Most seductive, it brought with it incalculable wealth—multi-trillion dollar markets that capitalistic society had never known before.

In return, the virus asked that those who participated in its "shell game" forgo any generative thinking. "Outside the box" became an expression defining the forbidden territories of generative thoughts.

Entrepreneurial Generativity

Perhaps the most powerful principles that we conveyed to the leaders of the insurance industry in Springfield were those of "entrepreneurial generativity." This was the real genius of the American capitalistic system. From our earliest days, we have been driven by the generativity engines of Franklin, Hamilton, and Washington. Entrepreneurial leaders such as Edison, Ford, and Rockefeller defined the core infrastructure of the 20th century—electricity, combustion engines, motor cars, highways, and energy. Tom Watson, Sr., generated the model for the Golden Age of American Business in the second half of the 20th century with systems and sales of business machines and computers.

This was a tradition carried on by J. Walter Reardon. For his graduate thesis, Reardon designed the architecture for transforming the Department of Public Relations at Mass Mutual into the Division of Corporate–Community Communications. This became the platform for "mutual growth" of both the private sector and the community.

Entrepreneurial growth in the community came slowly as entrepreneurism was often viewed as a "parasitic enemy" rather than a "synergistic contributor" to the welfare of the community. Today there are many entrepreneurial enterprises led by graduates of the HRD program. Sephus Jackson, former Minister of War for B.A.N.G., the black revolutionary group, has generated security businesses that now employ young people from the streets he once roamed. Another kind of minister, Bryan Robinson, Jr., expanded his community services as bishop of the Macedonian Church of God in Christ.

Generating Entrepreneurial Enterprises

For the individual in the organization, the learning journey culminates in business and industry. Business is the most complex level. There the marketplace is the teacher. We label its learnings "entrepreneurial enterprise" and define our people products operationally as **Prime Human Capital** or operationally, **Entrepreneurial Enterprisers.**

When all components have stimulated and reinforced the citizens with these organizational operations at the highest levels, we may conclude that we have generated the **Organizational Capital Development** or **OCD** on which business operations can build.

Finally, the private sector partners with the public sector contributors in marketing operations built on organizational operations. They do so by employing **R^5 - Repositioning Processes** (see Figure 37):

- **Policy-Making** by Generating (G)
- **Executive Architecture** by Innovating (I)
- **Management Systems** by Commercializing (C)
- **Supervisor Objectives** by Commoditizing (C)
- **Delivery Tasks** by Attenuating (A)

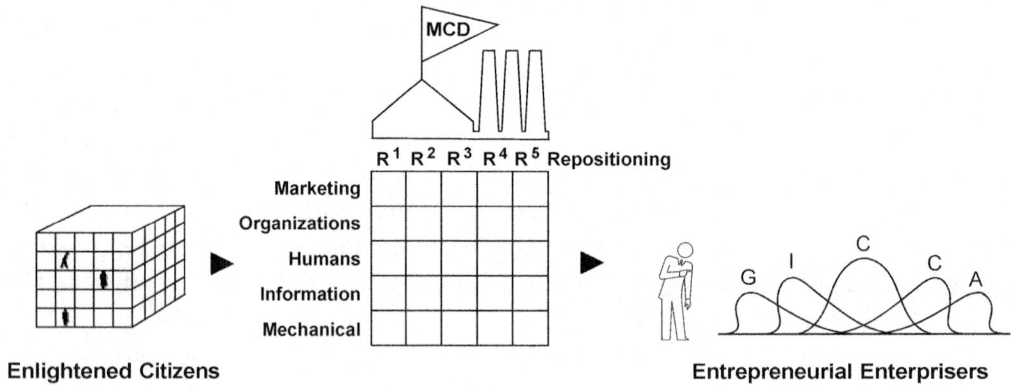

Figure 37. Building Business Operations

Cumulatively, then, the marketplace resources dedicated their **R^5 - Repositioning Systems** to Entrepreneurial Enterprise: after reasoning with images and reorganizing based on these images, people analyze the marketplace requirements and diagnose their organizational capacity to meet these requirements. For example, if the new requirements for organizational development are human-centric, then we must diagnose our organizational capacity for this dimension. This will involve multidimensional, curvilinear, and changing representations of the marketplace. Again, the repositioning objective is to maximize meeting both requirements and capacities at the highest levels.

In terms of the business and marketplace operations, then, enlightened citizens are transformed into entrepreneurial enterprisers with marketplace generativity skills. Once again, the generative process is a synergistic relationship among requirements, capacities, and placement in the market: each grows in relation to the other's growth. This is what makes Human and Organizational Capital into Marketplace Capital.

Perhaps the most dramatic examples of outgrowths of business generativity were entrepreneurial businesses serving the community-at-large:

- **HRD Press, Inc.—third leading publisher of training materials for Human Capital Development**
- **Human Technology, Inc.—the highest-rated business for Human Capital Development in the federal sector**
- **GenStar, LLC—the R&D extensions of Organizational Capital Development exploding on the software marketplace today**

All of these are demonstrations of the successful applications and transfers of human and information technologies began in the business generativity of the Springfield project. These and other businesses contributed to the continuing generativity of New Capital Development: Human, Information, Organization, Marketplace, Community, Cultural, and Economic Capital Development.

Community Capital Development

In this context, the Generative Community is the source of Community Capital (see Figure 38). As may be viewed, every component in the community is the developmental source of New Capital.

In this cumulative manner, the community capital builds the foundation for cultural change and cultural capital:

Mechanically Empowered Children

Discriminative Learners

Generative Thinkers

Enlightened Citizens

Entrepreneurial Workers

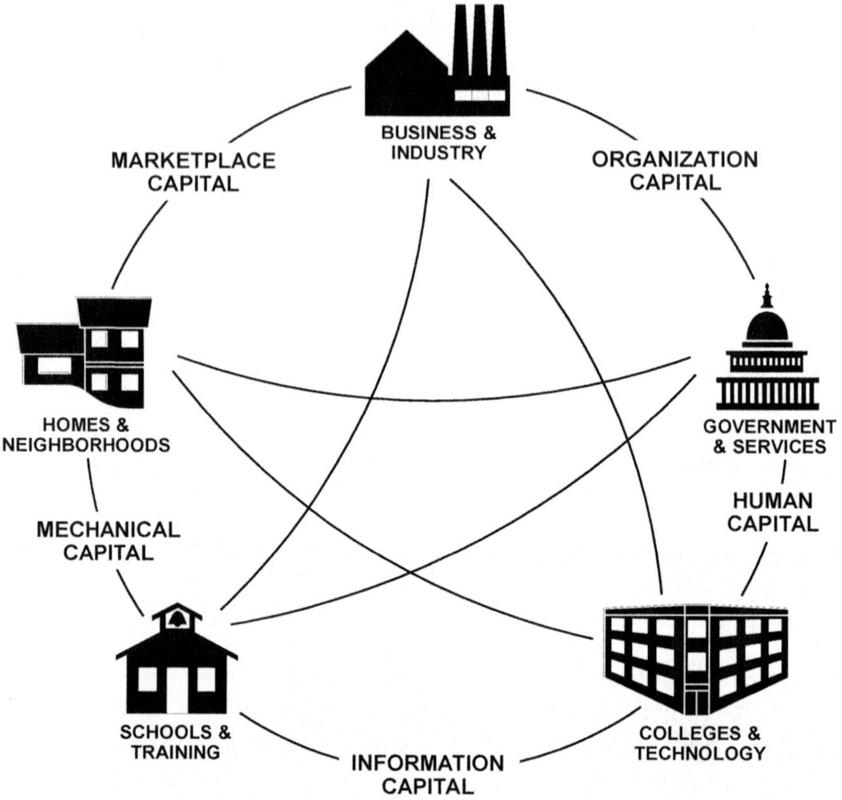

Figure 38. Community Capital Development*

These potent sources of growth operate synergistically to generate the truly capital community, a prepotent source of the "Generativity Solution."

Empowered by the Community Resource Development Programs of the 1960s and 1970s, the results were identified as "The Springfield Miracle!" The historic "City of Homes" had been transformed from the dying industrial town of the 1960s to the thriving Information Technology Center of the 1980s:

> **"Today, the city's industrial plants have been shutting down, but service industries and small-scale 'job shops' have taken their place. The biggest growth has been in the financial services, accounting firms, and insurance companies. All of these factors have converged to reduce unemployment below 3 percent."****

---

* The Entrepreneurial Community is available for viewing in animation on Pod V of **mcleanproject.com**

** Nocera, J. The Springfield Miracle. *Newsweek,* June 6, 1988, pp. 45–48.

In our graduate and undergraduate programs at American International College, the dying industrial town had discovered that it was mismatched to meet the conditions of the evolving Information Age.

Through generative leadership, its community components were aligned to facilitate the instantaneous communication of growth-relevant information. All community components were integrated in community generativity. For example, the minority groups actually reached the standards of performance of the majority groups on indices such as earning power.

This was a glorious time for Springfield and its citizens! Empowered by the "Skyhooks of Generativity," the people had pulled themselves up by their own "bootstraps of growth"—personal, educational, governmental, and business in "The Synergy of Corporate–Community Growth" modeled in the graduate program by Reardon's architecture.

Springfield had risen—like Phoenix out of the ashes—to define itself as a thriving "Information Center." It had reinvented itself as a generative community. It had generated its own "Generativity Solution."

- **In total, the business generativity initiated originally in Springfield contributed substantially to the creation of HCD market greater than one trillion dollars.**
- **Of this market, the R&D of two corporations alone generated an income of over one-half billion dollars.**

Part VI

Cultural Generativity

17

From Generativity to Degenerativity

The greatest impact on American generativity and innovation was the shift from a producer culture to a consumer culture. The producer-driven economy generated five decades of incredible economic growth. In turn, when converted, the consumer-driven economy led to the current economic bust.

How this came about is a complex tale of political economics. Why it came about is the more profound issue for it had disastrous impact on the American economy and its people.

Recall that America was rolling to victory in Europe and Asia when in 1944 the first Bretton Woods Conference was held. The result of this conference was to establish the American currency as the universal standard for all currencies. The dollar, in turn, was backed by its own standard, "The Gold Standard."

The Generativity Curve

World War II had unleashed the awesome productivity of the American people. Indeed, as much as anything else, America's soldiers were successful because of America's accelerating manufacturing capacity that provided the armaments and logistical support to the fighting men. It was a tribute to this very same morale, motivation, and mentality that warranted the Bretton Woods I concession. It tapped into the work ethic and fighting spirit of American men and women, many of whom were first generation Americans.

Indeed, it defined and reinforced a producer-driven economy that out-produced the world, not just in goods and services but in creativity and solutions. The American system was a "groove" for the producer mentality. In the 1950s, companies like IBM exemplified "The Golden Age of Industry" with "The GIC System":

- **Generators** made "breakthroughs" at the Watson Center.
- **Innovators** made "applications" at Advanced Systems Development.
- **Commercializers** made profits with their ubiquitous marketing arms.

The Generators, Innovators, and Commercializers related together interdependently in "The GIC System." Each component had its own unique functions, yet all components were committed to integrating their individual contributions into the company mission. Their motto was "THINK!" Recognizing that all stood together or all fell apart, they prospered as the leaders of "The Golden Age of Business for America."

As may be viewed in Figure 39, the producer-driven economy, led by Generators and Innovators, led the market on an elevating and escalating curve of growth. The people prospered and, indeed, flourished in generating their own unique American culture exemplified by the following conditions:

- **Cultural relating** transformed America from a "boiling cauldron" to a "melting pot" and provided the human capital support for "The Production Machine."
- **Participative governance** transformed Americans from the dependent minions of autocratic Anglophiles to the enlightened citizens of a transparent and accountable government.
- **Free enterprise** economics transformed Americans from extensions of their own industrial machinery to entrepreneurial capitalists in a Generative economy of their own making.

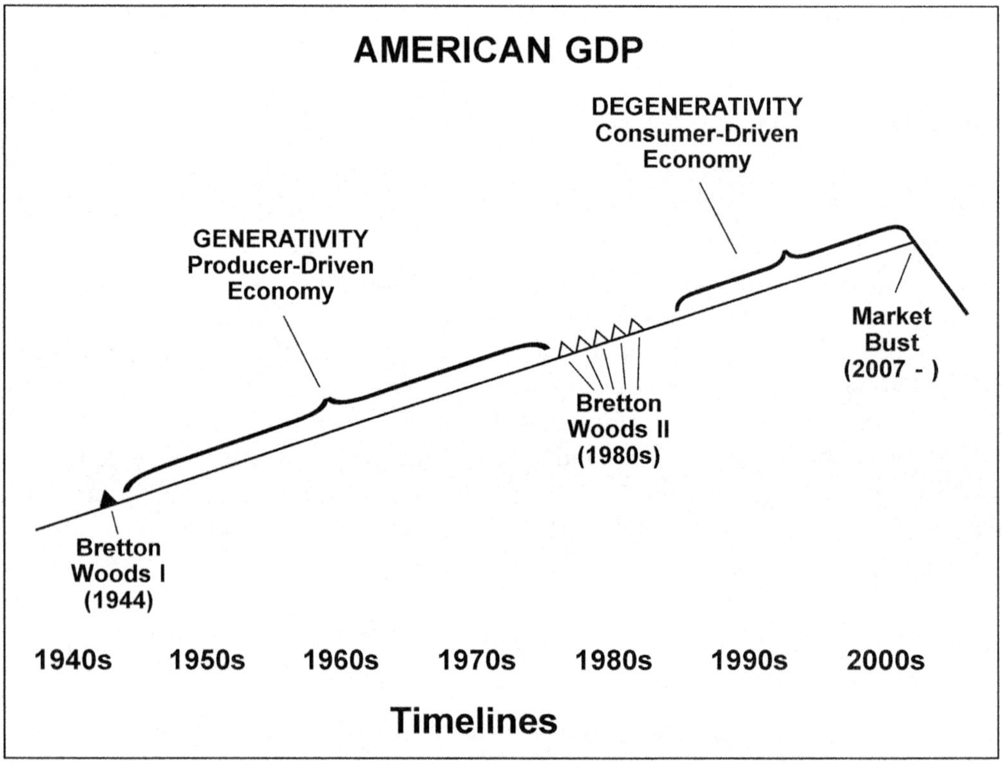

Figure 39. The Effects of Interventions Upon Generativity

Degenerativity

Things were looking great when some of America's alleged leaders overstepped their political boundaries. During the 1980s, in their conflict with the Soviet Union for world domination, they began to negotiate deals with nations around the world. In the heat of passion, Kissinger's pragmatism replaced Reagan's idealism: in so doing, it transformed Reagan's divine vision of "the city on the hill" into "a little for you and a little for me" diplomacy:

> **Adopt democratic governance and defeat communism and you will grow rich as our trading partners.**

There was only one problem—a huge problem! The countries of the world wanted to trade only if they had a positive trade balance. It was left to America to choose: between demonstrating and dominating; between leading and saving the world; between capitalism and ideology; between prosperity and "faux prosperity." America chose poorly!

In the 1980s and early 1990s, there were a series of conferences between America and its leading trading partners. Informally, these meetings may be labeled "Bretton Woods II," although American leaders ignored the same ideology they were exporting: the American citizenry did not participate in the nontransparent decision making; the American people did not contribute to defining "The Free Enterprise Vision."

The sum of Bretton Woods II was this: America continues to trade at a deficit, and its trading partners initiate to buy their debtor bonds. Initially, Japan and now China have assumed the role of creditor bond-buying nations. "It was a good deal for the peasants," our Ivy-League leaders concluded.

The net of the initiatives of Bretton Woods II was to transform American culture from a producer culture to a consumer culture. The "good deal" was that Americans could purchase things at the cheapest prices in the world and consume more than any people in the history of the world. All questions were answered with the basic Walmart question: "Do you like the prices, or what?"

The long-term problem was this: when we changed the culture, we changed the character of the American people. The older generations were confused and disoriented. Were we supposed to model ourselves after scientists, teachers, and mechanics who pursued knowledge and empowered expertise? Or were we supposed to model after businessmen and traders who pursued profits and produced cheap goods? The younger generations were to answer these questions.

Almost instantaneously, the younger generations were impacted by the "one-trial conditioning" of the ubiquitous media campaign: "Work Less and Buy Cheap!" They bought whole hog the assumptions of consumerism as the driving force in the marketplace. Indeed, they made it into the ethic of their generation. To be sure, they labeled it "The Consumer Ethic."

Some of us in the older generations lost contact with this new ethic. We thought that we had to learn some substance and produce something of value. We thought that we had to demonstrate that the way of American productivity was the best way: the healthiest way, the most civilized way, the most moral way in the sense of the Judaic Christian morality upon which this country was built. We believed that we would lead the world by producing—not by consuming!

Like "a shooting guard without a conscience," the younger generations applied the ethic "Buy!" It was the answer to every market explosion: "Buy!" Indeed, it was the answer that Bush II gave to the market depression: "Buy!" To be sure, it is the answer we hear to our current economic depression: "Give them welfare money so they can buy!"

It is the death gurgle of a dying civilization: "Buy!"

We desperately need a Bretton Woods III: to install the producer engine of economic growth; to drive consuming by producing; to generate possibilities and satiate probabilities.

18

The American Culture Shock

On June 14, 2004, I addressed the first "Freedom-Building Conference" in McLean, Virginia. I remember it well because Flag Day is my birthday. At that time I expressed my disappointing projections of the Bush administration's functioning on the empirically validated "Three Factors of the Freedom Experience":

- On cultural relating, I projected a disastrous decline from collaborative relating to dependency reactivity.
- On participative governance, I projected a precipitous plunge from democratic participation to totalitarian socialism.
- On free enterprise economics, I projected a desperate fall from entrepreneurial capitalism to a command-and-control economy.

To my dismay, all of this has to come to pass.

Perhaps the greatest culture shock in the last 100 years of America was experienced by its citizens in the presidential term of George W. Bush. From an administration that began with the mission of "compassionate conservatism," this administration devolved to one that emphasized "passionate despotism."

Bush's "compassionate conservatism" was to be found in the following "Pre-9/11" goals reflecting his principles:

- Improving the lot of the underprivileged
- Improving our international relations
- Reducing government, taxes, and bureaucracy
- Transforming our national defense structure
- Engaging in bipartisanship, which had characterized his governorship in Texas

Bush was to abandon all of these goals.

"Post-9/11" changed everything. Bush had become the "wartime president":

- His flagship legislation, "No Child Left Behind," proved a sham in practice.
- His humility in international relations succumbed to the totalitarianism of the neo-conservatives.
- His libertarian intentions were deformed by profligacy in spending, which resulted in larger government and massive debt and deficit.

- His commitment to security was undermined by a "hard-wired bureaucracy" that could not manage the extraordinary defense budgets it received.
- His bipartisan humanity produced governance bitterly divided by partisan politics and petty prejudices.

Bush failed on all of his goals.

We, the American citizens, suffered the consequences. When we analyze the characteristics of the "American Experience" that made us peaceful, participative, and prosperous, we conclude the following about Bush's administration:

- He undermined the uniquely American entrepreneurial capitalism that made us prosperous by conducting a totalitarian socialistic government in the service of the multinational corporations that he believed he required in his capacity as "wartime president."
- He eliminated the uniquely American "balance of powers" that enabled a participative democracy by ruling by edict and intimidation, assuming the powers of Congress, and politicizing the decisions of the Supreme Court.
- Perhaps most disastrously, he transformed our uniquely American "melting pot of cultural relations" back again into a "boiling cauldron" by pouring salt on the wounds of ancient prejudices and hatreds with "divide and conquer" strategies.

Yes, we American citizens know first hand what "culture shock" is all about!

19

Cultural Capital Development

In defining the operations of cultural generativity, we have factored the values centering on cultural capital development: cultural relating, participative governance, free enterprise. Together, these values define the Freedom Functions. They emphasize the levels of freedom that cultures must achieve in order to meet the requirements for participation, contribution, and leadership in the global village and its marketplace (see Table 3).

Table 3. The Cultural Generativity Architecture

		THE CULTURE FUNCTIONS			
	STAGES OF CCD '	**Cultural Relating**	**Participative Governance**	**Economic Enterprise**	
GROWTH ↑	5. Leader	Free Independent	Free Enterprise	Free Democratic	21st Century
	4. Contributor	Collaborative	Capitalistic	Representative	Late 20th Century
	3. Participant	Independent	Mixed	Mixed	Mid 20th Century
	2. Observer	Competitive	Command	Authoritarian	Early 20th Century
SURVIVAL	1. Detractor	Dependent	Control	Totalitarian	Pre-20th Century

As may be noted, the culture functions define the dimensions of cultural freedom for nations:

- **Cultural relating** or the degrees of relatedness within, between, and among cultures
- **Participative governance** or the degrees of participation in governance within, between, and among cultures

- **Free enterprise** or the degrees of freedom in economic enterprise within, between, and among cultures

Together, the cultural functions define the cultural architecture for entry and elevation in the 21st century global marketplace.

20

Freedom Functions

Cultural Relating

With cultural relating, all human endeavors are probable with a 95-percent degree of certitude. Specifically, with cultural relating at high levels, all national initiatives are possible with a similarly high degree of certainty. The plain fact of the matter is that nations cannot trade if they do not relate. Empathically, nations cannot govern if they do not relate. Emphatically, nations cannot participate in the global village and its marketplace without relating.

In this context, the levels of cultural relating are defined in the Cultural Capital Development Model or CCD′ (see Table 4):

Level 5: Free and interdependent relating is defined by "mutual processing for mutual benefit" within, between, and among other cultures and/or nations.

Level 4: Collaborative relating is defined by shared goals and common means to achieve the goals within, between, and among other cultures and/or nations.

Level 3: Independent relating is defined by reserving the right and power to compete as well as cooperate within, between, and among other cultures and/or nations.

Level 2: Competitive relating is defined by achieving and maintaining the "competitive edge" within, between, and among other cultures and nations.

Level 1: Dependent relating is defined by dependency in relating within, between, and among other cultures and nations, including adapting and reacting.

Table 4. Cultural Relating Scale

LEVELS	CONCEPTUAL DEFINITIONS	BEHAVIORAL OPERATIONS
5. Interdependency	**"Merge & Go"**	
4. Collaboration	**"Get, Give, & Merge"**	
3. Independency	**"Get & Give"**	
2. Competitiveness	**"Give & Go"**	
1. Dependency	**"Go"**	

Participative Governance

In the enlightened, modern world, governance is defined by participation. Government without participation restricts decision making to a council or a person of increasingly isolated, totalitarian "mind-sets": they or he make decisions about citizens for citizens. And with the acceleration of change, they "know" increasingly less and less, until they "know" nothing about anything! Even when they are overthrown or replaced, one limited set of variability is replaced by another limited set of variability. The only real discrimination in governance is the degree of participation.

In this context, the levels of participative governance are defined in the CCD' Model (see Table 5):

Level 5: Free and interdependent democratic governance defined by direct and mutually beneficial relations between citizens and decision makers

Level 4: Representative democratic governance defined by citizen relations with representatives who represent them as decision makers

Level 3: Mixed governance as defined by occasional or intermittently democratic and authoritarian activities

Level 2: Authoritarian governance as defined by directed governmental edicts

Level 1: Totalitarian governance as defined by total control of political practices

Table 5. Participative Governance Scale

LEVELS	CONCEPTUAL DEFINITIONS	BEHAVIORAL OPERATIONS
5. Enlightenment	"Merge & Go"	
4. Representation	"Get, Give, & Merge"	
3. Independency	"Get & Give"	
2. Authoritarian	"Give & Go"	
1. Totalitarian	"Go"	

Free Enterprise

The key to the highest levels of economic enterprise is the modifier "free." It is not the trade that defines it. It is the free thinking that generates entrepreneurial initiatives to meet consumer needs that defines it. Specialty products make trade attractive. Trade agreements without attractive products are like rental properties without occupants. The plain fact of the matter is that nations cannot trade if they do not have something of value. And they will not have something of value if they cannot generate substance!

In this context, the levels of free enterprise are defined in the CCD′ Model (see Table 6):

Level 5: Free and interdependent entrepreneurially driven enterprise defined by "mutual trading for mutual benefit" with other nations

Level 4: Capitalistic economics defined by maximizing financial capital-based investments and minimizing governmental burdens

Level 3: Mixed economics defined by occasional or intermittently capitalistic and command practices

Level 2: Command economics defined by directed governmental command practices

Level 1: Control economics defined by total governmental control of economic practices

Table 6. Free Enterprise Scale

LEVELS	CONCEPTUAL DEFINITIONS	BEHAVIORAL OPERATIONS
5. Entrepreneurial	**"Merge & Go"**	
4. Capitalistic	**"Get, Give, & Merge"**	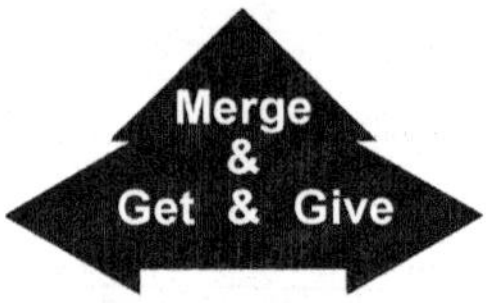
3. Independency	**"Get & Give"**	
2. Command	**"Give & Go"**	
1. Control	**"Go"**	

21

USA 2000—The Producer Culture

Over the last century, the U.S. defined the emerging requirements of the global marketplace:

- Free cultural relating
- Free participative governance
- Free enterprise economics

These are the requirements for entry into the 21st century marketplace. They also define the American Mission. Historically, this is how the U.S. was rated by subject-matter experts in 2000.

Collaborative Relating

Simply stated, a nation must relate culturally—within, between, and among cultures—in order to join the global marketplace. American leaders have always set their goals high on cultural relating (see Figure 40). Even transitionally when they retreated to independency or competitiveness, they never surrendered the ideas of collaborative relating. While their definitions of interdependency may have been limited, they were nevertheless striving toward it.

In 2000, the U.S. was rated at level 4.0 on cultural relating, reflecting its commitment to collaboration without real motivation for interdependency (perhaps as a function of the conflict of individualism with consensus), except within rare homogeneous cultural groupings.

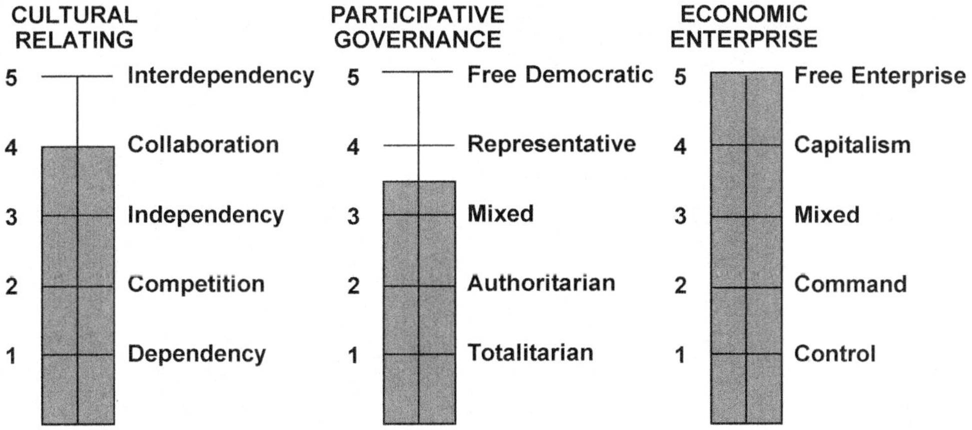

Figure 40. U.S. Levels of the Freedom Functions (2000)

Mixed Governance

The U.S. has also been dedicated to participative governance at high levels. While its citizens are guaranteed prepotent policymaking functions by the U.S. Constitution, its representative democracy often migrates into a "role reversal" where citizens serve politicians and bureaucrats. Nowhere has American promotion of democracy been more problematic than in attempts to impose it on foreign cultures. Participating begins with relating to others' frames of reference before empowering them to achieve their own unique purposes. That is the true nature of democracy. There is no freedom without empowerment! There is no empowerment without relating!

In 2000, the U.S. was rated at level 3.5 on participative governance (see again Figure 40), reflecting the inconsistent nature of its attempts at making representative governance work; stumbling that led precipitously to representative self-aggrandizement; stumbling that yields disastrously to authoritarianism by default.

Free Enterprise

The U.S. has been dedicated to free enterprise at the highest levels. Succinctly, the U.S. Constitution, protecting private property and, in an amended form, civil and economic liberties, provides a strong foundation for the nation's free and dynamic economy. In addition, the U.S. has taken the leadership in promoting free trade, not only with neighbors in the Western Hemisphere but with nations around the world. While not yet having realized its mission of universal free enterprise, the U.S. has been the world leader, par excellence.

In 2000, the U.S. was rated at level 5.0 on free enterprise, reflecting its aspirations for reciprocal trade at the highest levels (see again Figure 40). "The Producer Culture" continued on its path in spite of the influence of "The Consumer-Driven Culture." Whatever the obstacles, there would be no vision of the global marketplace without America's highly initiative leadership; there would be no standards for prosperity and, thus, for the evolution of world civilization!

To sum, in 2000, the U.S. received relatively high ratings on the freedom functions:

- Collaborative relating
- Mixed representative and authoritarian governance
- Free enterprise economics

The good news was that, however "ugly" externally, the U.S. continued to relate collaboratively internally and with friends and neighbors. Moreover, the U.S. was convergent externally and internally with its path-finding free enterprise mission. The bad news was protecting a democratic image in an authoritarian manner externally while meaningful participative governance was dissembling internally.

22

USA 2008—The Consumer Culture

Once again, over the last century, the U.S. defined the emerging requirements of the global marketplace:

- Free cultural relating
- Free participative governance
- Free enterprise economics

These are the requirements for entry into the 21st century marketplace. They also define the American mission. Currently, the same subject-matter experts gave very different ratings in 2008.

Competitive Relating

In 2008, the U.S. is rated at level 2.0 on cultural relating (see Figure 41). This reflects her growing suspicion and hostility following the 9/11 terrorism attacks and her inability to translate "spy in the sky" intelligence data into meaningful "feet on the ground" information upon which she could base strong and operational security systems. In the absence of operational thinking systems, America writhed in agony and engaged in random, non-responsive, and inappropriate initiatives emphasizing cultural competitiveness and the resultant cultural dependency reactions.

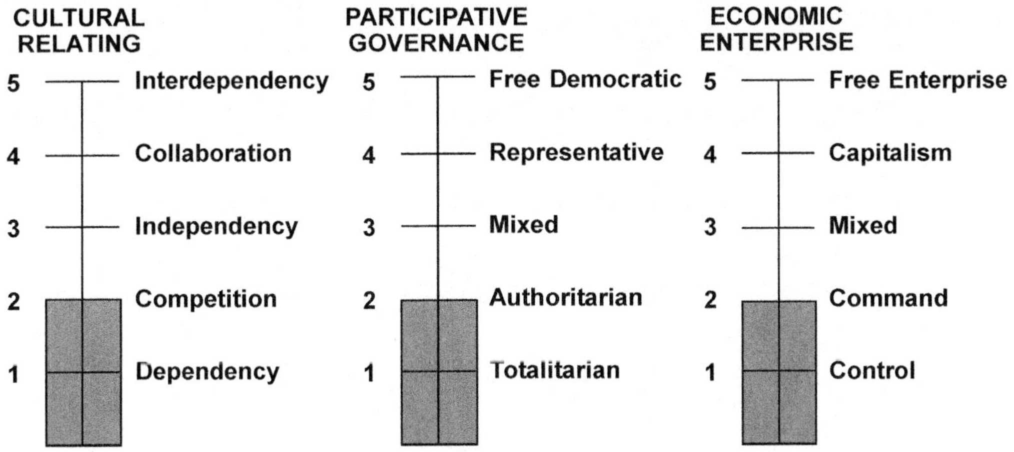

Figure 41. U.S. Levels of the Freedom Functions (2008)

Authoritarian Governance

In 2008, the U.S. is rated at level 2.0 on participative governance (see again Figure 40), reflecting the increasing totalitarianism of the party in power and the reactive authoritarianism of the opposition: rather than being consulted as the constitutionally mandated policy makers, We the People, the citizens, feel imposed on by the authoritarian inquisition: Either ______ or ______!

"Either you are a patriot or you're disloyal!"

Command Economics

In 2008, the U.S. is rated at level 2.0 on free enterprise (see again Figure 40), reflecting its shift in government sponsorship from a vision of an entrepreneurial-driven free enterprise system to government-reinforced multinational corporations that by their very nature conduct "command-and-control economics." In spite of their origins and support from American freedom functions, the totalitarian multinationals relate competitively and attempt to reduce their markets to dependency on "command-and-control strategies."

In summary, it is clear that America had indeed taken a "U-turn on the freedom road." America received low ratings on cultural relating, free enterprise, and participative governance. Under the administration of Bush II, the newly minted **"Government-Reinforced Industrial-Military" Complex** or **"GRIM Complex"** has placed America's citizenry as well as American destiny at risk.

- Competitive and dependent cultural relating placed peace at risk—internally within the nation and externally in the global village!
- Authoritarian and totalitarian governance placed participation at risk—internally within the nation and externally in the global courts of law and justice!
- Command-and-control economics placed prosperity at risk—internally within the nation and externally in the global marketplace!

In transition, peace, participation, and prosperity are the historical functions of "The Producer Culture." They are now placed at risk by "The Consumer-Driven Culture," which now accounts for 70% of GDP.

Part VII

Economic Generativity

23

The Generativity Cycle

For over 30 years, the "entrepreneurial capital" of the world has thrived in Silicon Valley, California, with the "Generation–Innovation–Venture Capital–IPO Cycle" generating new wealth, commercializing innovation, and creating new companies and industries while spinning off millions of new jobs. This "Cycle of Generativity" is self-perpetuating: entrepreneurs generate new ideas, form teams, innovate and plan, and promote the ideas to venture capitalists. In turn, "VCs" invest, "start-ups" perform, "IPOs" are offered. The cycle begins again with the profitability and reinvestment of the companies. The success of this engine in our economy is over, under assault by legislators and regulators who do not understand the fragile nature of generativity. In the name of "fairness," all governmental agencies have put generativity at risk.

These are the generativity functions in operation: the generator generates "breakthrough ideas"; the innovator transfers the ideas to the marketplace; the commercializer commercializes the ideas to maximize profitability. These are the critical phases of economic generativity, the culminating mission of the Generativity Cycle.

This, then, is the Generativity Cycle. Beginning with community generativity, individual, and organizational generativity are nested in the community, which is the generator of New Capital Development:

- **Marketplace capital** or positioning for comparative advantage
- **Organizational capital** or productive alignment with positioning
- **Human capital** or generative processing to implement alignment
- **Information capital** or information modeling to implement processing
- **Mechanical capital** or mechanical tooling to implement modeling

Together, these new capital ingredients provide the building blocks for cultural generativity.

In turn, cultural generativity is empowered by networks of generative communities capable of generating new capital. What holds these communities together are the freedom functions that culminate in cultural capital development:

- **Cultural relating,** which relates the communities collaboratively and interdependently
- **Participative governance,** which empowers the communities to democratic and enlightened governance

- **Free enterprise,** which empowers the communities to entrepreneurial and capitalistic enterprise

Together, the freedom functions provide the foundation for the spiraling benefits of "The Pillars of Civilization"—peace, participation, and prosperity.

Finally, economic generativity is found by networks of generative cultures. Economic generativity culminates in economic capital development:

- **Generativity,** which creates breakthrough ideas and architectures
- **Innovation,** which transfers these architectures into products and services dedicated to marketplace functions
- **Commercialization,** which markets and sells the products and services, yielding productivity and profitability in the marketplace

Together, these economic functions define the mission of a continuously changing and growing marketplace.

In sharp relief, then, the mission of economic generativity is very simple:

> **Economic functions are achieved by cultural components that are empowered by community processes.**

Together, all of these dimensions process interdependently and synergistically to generate explosions of wealth heretofore unknown to humankind.

The only real obstacles are those brought about by the reinforcement schedules of our consumer-driven culture:

- We have been positively rewarded to expect comfort and even luxury in our buying habits.
- We have been extinguished by neutral responses to our working habits.
- We have been negatively reinforced by punishments in our "producing initiatives."

In other words, our consumer-driven cultures have inverted the reinforcement schedules for an economic recovery.

Clearly, we can already see that the socialistic system undermines the operation of the Generativity Cycle. After an average of hundreds of individual public offerings (IPOs) per year over the past two decades, there have been only six IPOs from Silicon Valley in 2008, or 2% of the filings on an average year. Our choices are clear:

- Assume a static marketplace and "teach our children Mandarin."
- Assume a changing marketplace and "teach our children generativity!"

24

The Generativity Curve

The probabilities scientists have given us normal curves: nice, linear, idealized symmetrical distributions around central tendencies such as means or averages (see Figure 42). The curves tell us something about where people and products stand in relation to this central tendency. The work of the 20th century has been based on these probabilistic assumptions. The mission of the 20th century has been to move toward reducing our deviation from this idealized central tendency: in other words, to look more and more like the middle of the curve.

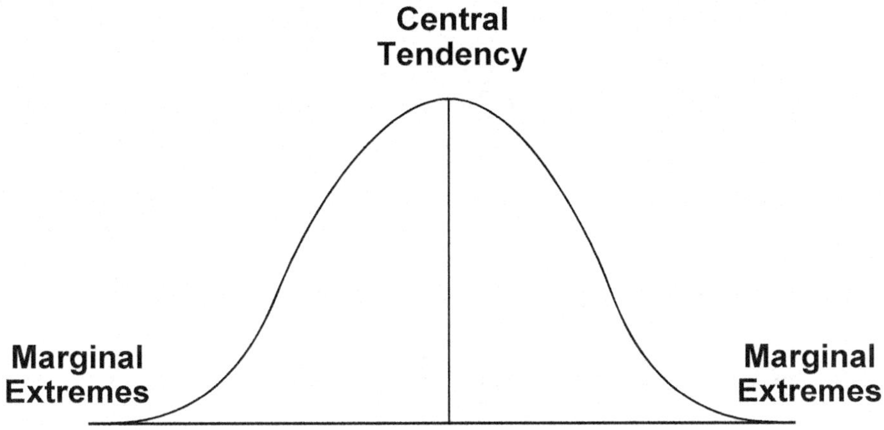

Figure 42. The Normal Curve

The normal curves also attempt to tell us how people and products move through the phases of the curve. For example, the normal curve has been adopted by the marketplace to reflect the movement of products and people through market phases: opinion leaders, early adopters, middle adopters, late adopters, and laggards or non-adopters (see Figure 43). While this approach has produced an extraordinary array of products and services, it has simultaneously generated apparently finite images of people who look, act, and think alike.

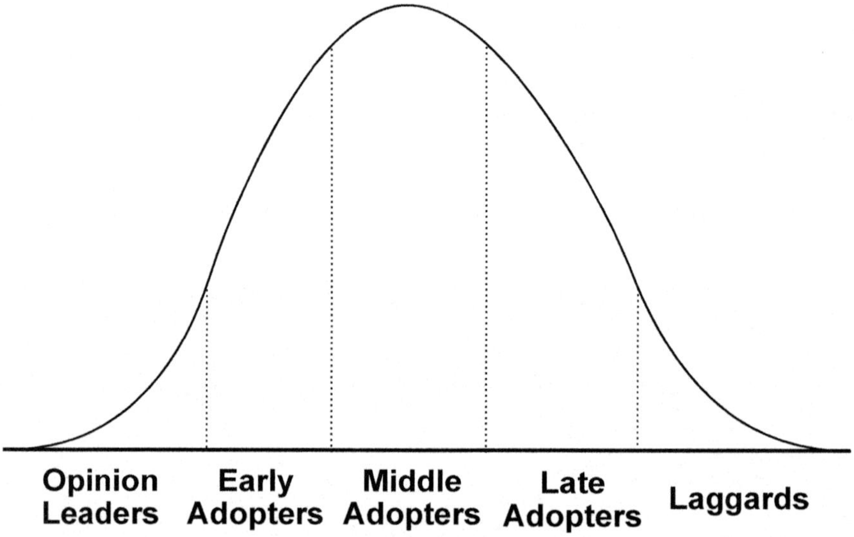

Figure 43. The Market Curve

The Market Curves

In fact, the market curve is not a normal curve. Indeed, nothing in God's universes is linear, normal, and symmetrical. To be sure, the so-called market curve comprises multimodal market curves: each, in turn, is multidimensional, interdependent, and curvilinear as well as asymmetrical and changeable (see Figure 44). We may view a phenomenal representation of these market curves in the illustrations below:

- **Generating** or stimulating markets
- **Innovating** or initiating markets
- **Commercializing** or making markets
- **Commoditizing** or draining markets
- **Attenuating** or breaking markets

For our purposes, we label these market phases "GICCA." They have profound implications for their corporate cultures and their people and products.

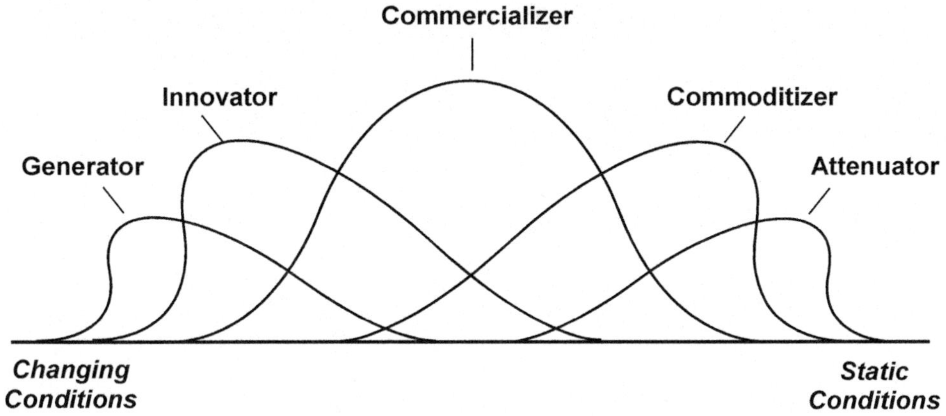

Figure 44. Multimodal Market Curves

The market curves are stimulated by the generator curve (see Figure 45). As may be viewed, this curve is tilted toward changing conditions and skewed away from static conditions. This means that the corporate culture and its people and the products are oriented to changing conditions, either by responding to existing needs or creating needs for future conditions. In this respect, the generator cultures are beyond any existing market curves and before any new markets. Indeed, they exist in a "limbo" of premarket conditions, often residing only in the minds of their inventors.

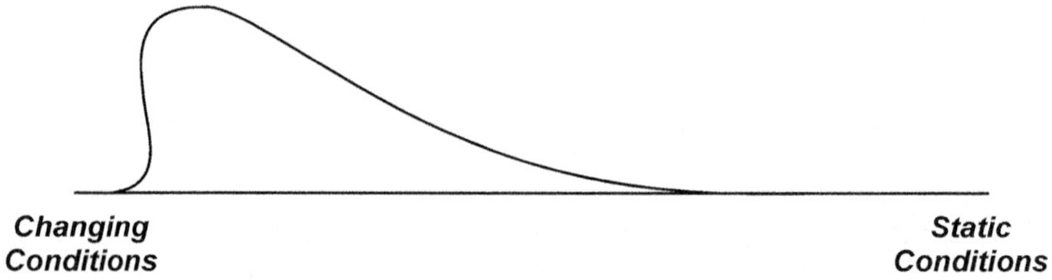

Figure 45. The Generator Curve

The real market curves do not come into being until the innovators step into the picture (see Figure 46). As may be viewed, the innovator curve is tilted toward the substance of the generator and the changing conditions of the marketplace. This means that the innovator culture has one eye on the ideas of the generator and the other eye on the potential of the marketplace. In this respect, the innovators initiate the market curves. More than anyone else, they are responsible for the creation of the commercial markets.

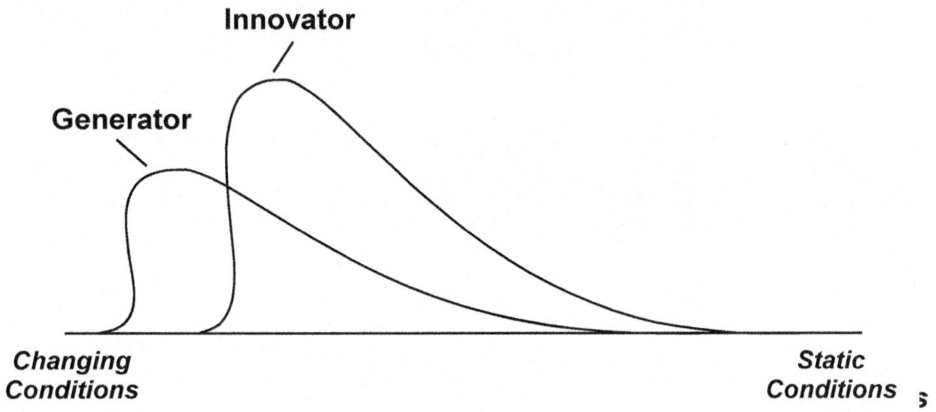

Figure 46. The Innovator Curve

The commercializers are, of course, those responsible for making and, then, "peaking" the market curves (see Figure 47). As may be viewed, the commercializer curve is directed straight up to respond to customer needs in order to achieve the indices of robust sales. This means that the commercializer cultures have both eyes on maximizing the indices of market commerciality: productivity, gross, profit, growth, market share. In this respect, the commercializers *are* the commercial market: they produce more so that they can profit more!

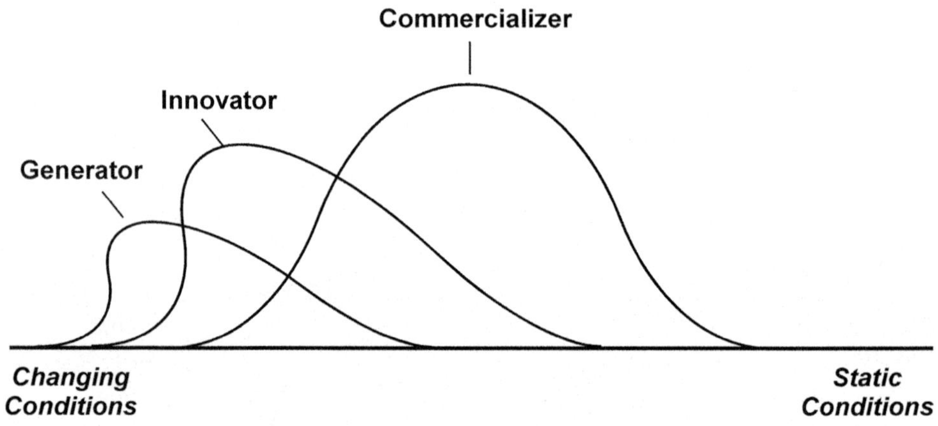

Figure 47. The Commercializer Curve

The appearance of the commoditizers are early warnings of the decline of the commercial market (see Figure 48). Indeed, often the commoditizers cause the decline of the commercial market: they produce more for less! As may be viewed, the commoditizer curve is tilted toward the static conditions of the marketplace and away from changing conditions. As such, the commoditizer cultures seek the security of expanding sales even in the face of declining profits.

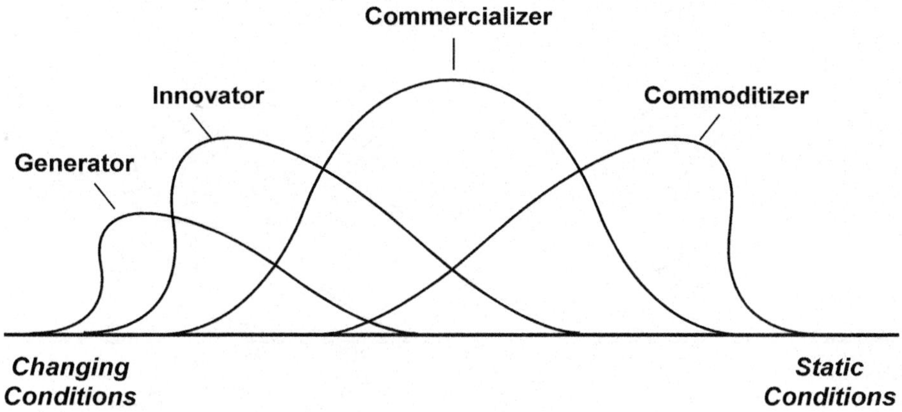

Figure 48. The Commoditizer Curve

Ultimately, the attenuators take over this decline (see Figure 49). As may be viewed, this curve is tilted toward stability and skewed away from change. The attenuator curve is a curve of attenuating benefits: producing more and more for less and less. The apparent stability sought by the attenuator cultures is undermined by the ultimate reductionistic principle of attenuation: producing everything for nothing! For our purposes, it completes the representation of the GICCA curve.

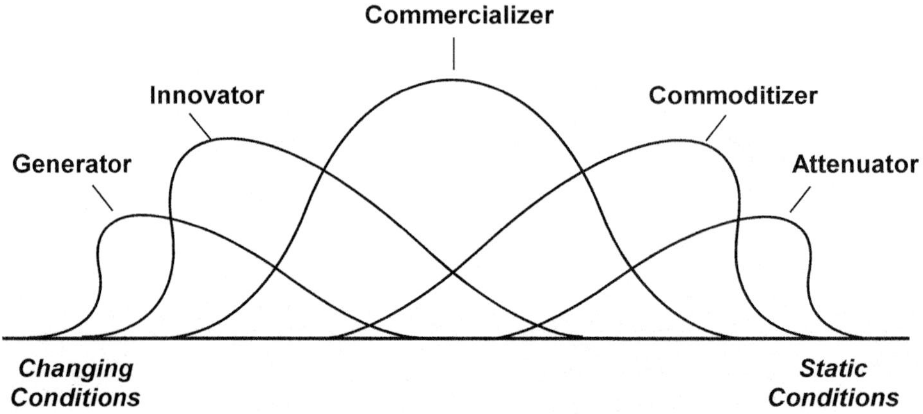

Figure 49. The Attenuator Curve

25

The Generators

The "generators" are a unique culture of personality types dedicated to responding to the changing conditions of the marketplace. In this context, the generators are dedicated to generating breakthroughs that impact the changing conditions: primarily in the form of the substance of science and technology. As such, the generator curve stimulates the market curve by generating ideas of the possible that are adopted by innovators of the probable. By way of illustration, we may study the evolution of the semiconductor or "microchip."

The evolution of transistors and semiconductors are inextricably bound together. It has been a little more than 70 years since Lee DeForest invented the vacuum tube. The vacuum tube, itself, was derived from the electrical operations of the lightbulb and then, in turn, was replaced by the transistors invented by John Bardeen and Walter Brattain based on a theoretical framework developed by William Shockley in 1939.

Unfortunately, transistors had to be hand-wired together on panels, with less than a 50% success rate on crude assembly lines. Nevertheless, the transistors were the obvious substitutes for vacuum tubes on awkward electronic computers such as ENIAC.

It was then that Robert Noyce began to work on assembling sheets of silicon or other materials without wires. At this point, Noyce began to detail designs for a solid-state circuit. The integrated circuit made it possible to create miniature computers to replace the monstrous computers then available. At some point, Noyce became aware of a separate and distinct effort being made by another engineer.

In 1959, Jack Kilby's idea was to integrate all the parts of an electronic circuit. He was successful in his efforts, eventually integrating the resistors, transistors, capacitors, and diodes in a single monolithic block of semiconductor material.

The principle was labeled the "Monolithic Idea" in honor of the monolithic integrated circuit. Today, the invention is better known as the "semiconductor chip." The economic implications of this technological innovation were—and still are—profound. Since then, the chip has pervaded our land, sea, and sky. It functions at the heart of our deep-space probes and our space shuttles. It is the core of our deep-sea sensors. On earth, it drives our clocks, cameras, toasters, word processors, and perhaps most critically, our communications networks.

The semiconductor chip has made possible astonishing reductions in the size, cost, and power requirements of electronic equipment. As a technological innovation, the chip is smaller, lighter, faster, cheaper to make, and more reliable than traditional circuits. Indeed, the chip's raw material, silicon, is derived from ordinary sand, one of earth's most plentiful elements. Finally, while the manufacture of microchips can be demanding, it does not pollute earth's land, sea, and sky.

The chip's most important implications may be human rather than simply technological. The chip has made computers almost universally available. A single chip, driven by inexpensive batteries, is faster, more reliable, and more powerful than the largest, most energy-consuming computer of the 1950s. In a very real sense, the microchip brought about the end of the Industrial Age and ushered in the Electronics or Data Revolution. Like all true revolutions, it introduced the prospect of a new age: the Information Age. In so doing, Kilby's work may be labeled "Extrapreneurial" to signify the generation of entirely new markets.

At this point, he begins to engage in generative processing:

- Establishing his goals or images of products or outcomes
- Exploring the current systems for achieving the outcomes
- Expanding alternative systems for achieving the outcomes
- Narrowing to new systems for achieving the outcomes
- Building new systems for achieving outcomes
- Feedbacking the productivity benefits for results benefits compared to resource investments of the outcomes

An illustration of the phases of generative processing may be viewed in Figure 50. As illustrated, the phases of processing occur as follows:

- **Pre-Processing:** Goaling by valuing future outcomes
- **Phase I:** Exploring by analyzing current operations
- **Phase II:** Understanding by synthesizing new operations
- **Phase III:** Acting by operationalizing new objectives and programs
- **Post-Processing:** Feedbacking by evaluating outcomes

Together, these phases constitute a processing system. Feedback from evaluating outcomes recycles the goaling and processing phases at higher and higher levels.

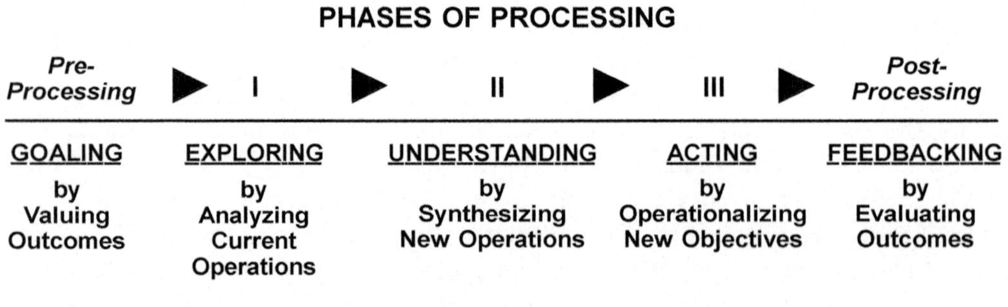

Figure 50. The Phases of Generative Processing

Moreover, once he has developed productive images, Kilby shares them with his colleagues. He stimulates them to generate and share their images. Together, they treat all images as input to interpersonal processing of new and more powerful images.

Finally, and most important, Kilby lives and processes interdependently inside the phenomenal images he and his colleagues have generated. He dedicates himself along with other phenomenal components to accomplishing the phenomenal functions. He employs the phenomenal processes to enable the components to accomplish the functions. In effect, he merges with the phenomena. What he has generated, he now innovates.

Kilby is the model of the generator. He builds models of the phenomenal operations. He processes these operations generatively—individually, interpersonally, and finally, interdependently.

Processing is all about phenomena. Phenomena are the people, data, and things that we encounter in our daily existence. Phenomena may be as small as a skill step in a task we must perform. They may be as large as the great mission of our business, our economy, or our society. Every experience is phenomenal.

Phenomena are defined by their operations: the functions they discharge; the components they invest; the processes in which they engage; the conditions from which they derive; the standards they set. In essence, phenomena are processing systems: component inputs are transformed into function outputs by processes under specifiable conditions and with measurable standards.

Generative processing, then, is about generating phenomena: qualitatively improved images of phenomenal functions, components, processes, conditions, and standards. Ultimately, we learn to live and process within the phenomena we have generated. Generative processing is dedicated to generating qualitatively better and more powerful images of the phenomena with which we work! In short, generative processing is a "best processes" approach to breakthrough thinking.

In our own R&D experience, we have been privileged to generate new models of science: "The New Science of Possibilities." Building on probabilities science, possibilities science is calculated to drive the old science. The possibilities science has empowered us to develop deductive models for all processing. For example, we may view the corporate positioning matrix illustrated in Figure 51. It gives us the opportunity to determine the levels of requirements and values to which we aspire. All positioning begins with requirements and values.

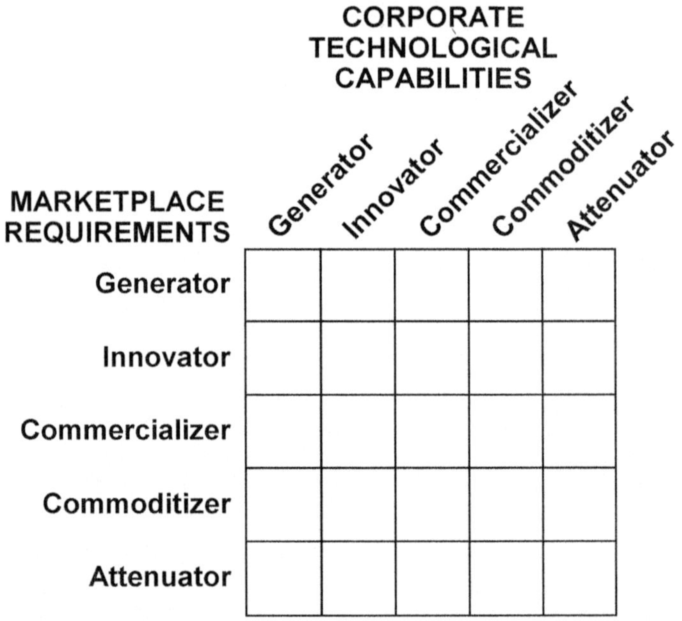

Figure 51. The Organizational Positioning Matrix

The positioning matrix enables us to determine the level of requirements to which we aspire as well as to determine the level of corporate technological capability that we have to achieve our requirements level. Clearly, for example, it is not enough to want to be known as technological leaders in the field: we must have the technological capability to achieve that leadership, just as Kilby did.

To be sure, Kilby became the model for both generator and extrapreneur in the latter half of the 20th century. Yet each and every one of us may become generators and extrapreneurs in the 21st century. It may be the only way to keep abreast of the spiraling changes of our time. It *is* the only way to make our contributions to these changes.

There are as many different kinds of cultural types as there are market curves. Most prominent among these cultural types are the generators who generate "breakthroughs" in thinking, marketwise or otherwise. Similarly, there are many characteristics that can be employed to define the generator culture: physical resources that energize, emotional resources that catalyze, intellectual resources that actualize their contributions to generating breakthroughs. Preeminent among the characteristics of the generators, however, are their relationships with the market curve as in Figure 52. As may be noted, the generators relate not only to changing conditions but also to the expansion of the market curve. In other words, the generators relate to help define the changing conditions of the marketplace.

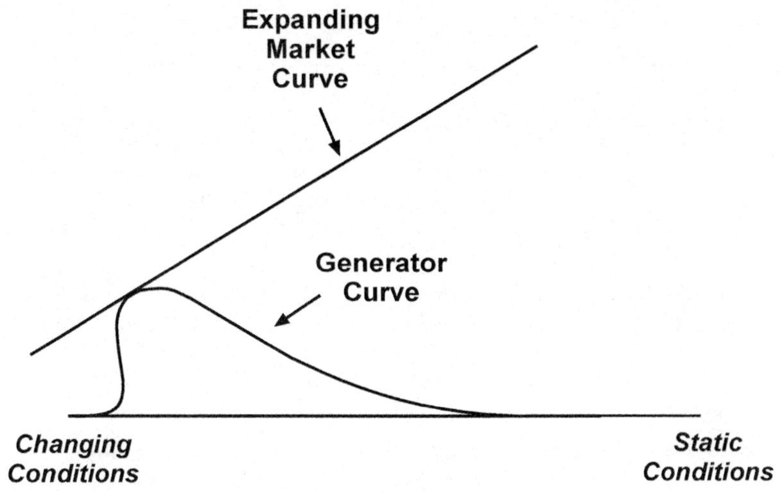

Figure 52. The Generator Curve

Culture is defined by its operations. Above all else, it is defined by the functions to which the other operations are dedicated. In this context, the generator culture may be defined as follows:

- Dedicated to breakthrough functions as the overriding mission
- Implemented by scientifically defined phenomena as the prepotent components
- Enabled by interdependent processing to continuously generate new images
- Housed in freeing contexts that, indeed, impose no "conditions" at all
- Assessed by continuously changing standards that generate futuristic operations

These, then, are the operations of the generator culture and personality type:

- Breakthrough functions
- Scientific components
- Interdependent processes
- Freeing conditions
- Continuously changing standards

Want to be a generator? Then you must pay the price of conquering substance that empowers you to be:

- Smart enough to know what breakthrough functions are
- Humble enough to embrace the phenomena scientifically
- Dedicated enough to process interdependently with everyone and everything
- Committed enough to understand that the price for freeing conditions is responsible for generativity
- Open enough to know that we are all responsible for the changeable standards of the phenomena with which we process, including our own

Want to develop a generator culture? If you can meet these requirements, then you can have these benefits. You will become a truly generative culture, actualizing all of God-given talents to enrich and empower our man-driven civilizations.

26

The Innovators

The "innovators" are also a unique culture of their own personality types. They respond to the changing conditions of the marketplace by transferring the breakthrough thinking of the generators. With one eye on technological substance and one on commerciality, the innovators are the true initiators of the expanding market curve. By way of illustration, we may continue our study of the evolution of the semiconductor or microchip.

The ranks of innovators are legion. It would not be exaggerating to say that the past 50 years have been an Age of Innovation. These are not the people who make the generative breakthroughs. These are the people who take the breakthroughs and transfer all or some of their operations into new products and, indeed, new services. To be sure, whereas the generators tend to be extrapreneurial in their initiatives, contributing to entire industries and markets with their work, the innovators tend to be entrapreneurial in their efforts, building new products and services for expanding businesses and industries.

For example, following in the wake of Noyce, Ted Hoff and a group of Intel engineers innovated Intel's gigantic microprocessor and thus that new business. Similarly, Steve Jobs and a group of Apple engineers innovated the Macintosh computer and thus expanded the computer market. Likewise, Don Estridge and a group of IBM engineers innovated the personal computer and thus expanded that business. Also, James Gosling, Patrick Nelson, and a group of Digital engineers innovated the Java computer language and expanded that business.

Indeed, the commonality of entrepreneurial initiative within innovative functions led to recommendations for corporations to devise new ways of cultivating entrepreneurial efforts to keep innovation within the corporation in order to capitalize on it.

Entrepreneurial initiative is not the only common characteristic that innovators have. Other operations are more basic:

- First, innovators are uniformly committed to a common mission: the transfer of already existing operations of phenomena, especially products
- Second, innovators are universal in their dedication of multiple and multidimensional technological components to this transfer function
- Third, innovators work in teams and process interpersonally to develop "the best ideas" for enabling the technological components to accomplish the transfer function

- Fourth, while they do not require the full freedom of generators, innovators require the releasing conditions—intentional or inadvertent—which allow them to pursue their innovations
- Fifth, although they are not dedicated to continuous change standards as are the generators, innovators allow for changes in standards as business has run its course

Together, these operations define the innovator:

> **Transfer functions discharged by technological components enabled by interpersonal processing under releasing conditions and with change standards.**

Innovators, themselves, do not operate deductively to accomplish "the great ideas" of their generative counterparts. Nor are they motivated by larger motives such as making contributions to the elevation of civilization or science.

Instead, innovators are directed inductively and interpersonally toward "the best ideas." They are motivated by business reasons for "building a better mouse trap."

"The Transfer Matrix" illustrated in Figure 53 is an illustration of innovation in ideation. It does not represent the processing of great brainpower like that of Kilby and Noyce. However, it does represent the simple transfer operations in which innovators engage. It will also serve to introduce the reader to the transfer functions to which innovations are dedicated. In this context, we dedicate this innovation gift, The Transfer Matrix, to the illumination of all innovators and "wanna-be" innovators.

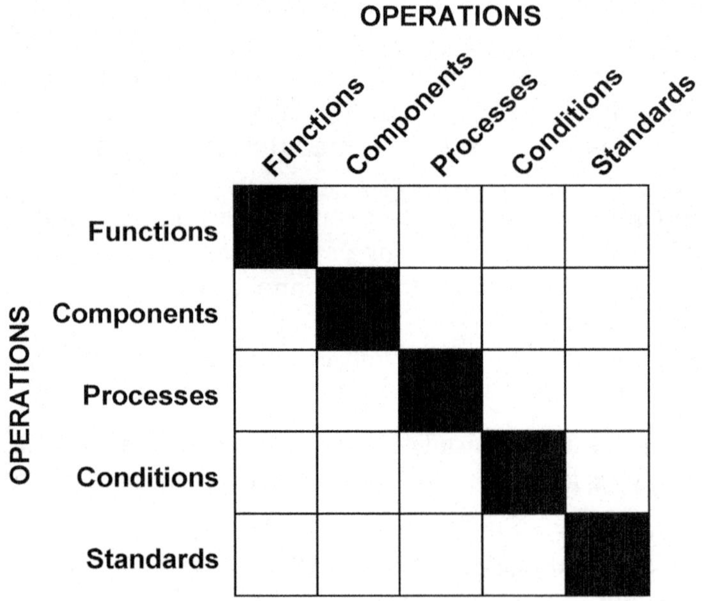

Figure 53. The Transfer Matrix

Here is how to use "The Transfer Matrix:"

1. Take any product, service, or business.
2. Define it by its operations: functions, components, processes, conditions, standards.
3. Align these operations on either the *x* or *y* axis.
4. In turn, hold each of these operations on one axis constant while you vary the other operations on the other axis.
5. In turn, vary each of these operations on one axis while you hold other operations of the other axis constant.

This series of exercises will yield transfer alternatives. In a typical innovation, the functions will be held constant while the components, processes, conditions, and standards are varied. This is the generative innovation that Noyce contributed months after Kilby invented the semiconductor chip.

Just as the generators, the innovators also relate to the expanding market curve (see Figure 54). Like the generators, this requires physical fitness that yields energy; emotional fitness that yields motivation; intellectual fitness that yields innovativeness. However, unlike the generators, the innovators do not stimulate the curve. Rather, the innovators initiate the curve based on the stimulation of the generators. In other words, the innovators relate more to the beginning of the curve than to the changing conditions of the marketplace. They do not so much define the conditions of the marketplace as to become part of them, to fill them out, to give them life and endurance.

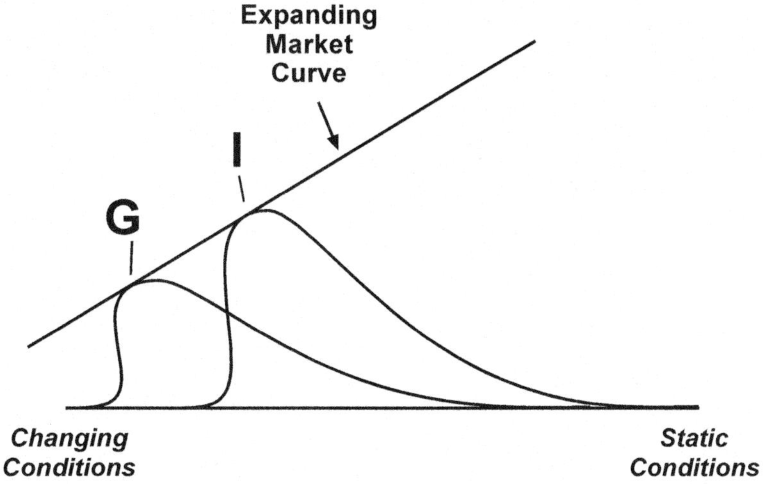

Figure 54. The Innovator Curve

In this context, the innovator culture may be defined as follows:

- Dedicated to the transfer of market breakthrough
- Implemented by technological components
- Enabled by interpersonal processing
- Nested in releasing conditions
- Measured by changeability standards

These, then, are the operations of the innovator in both market phase and cultural or personality type:

- Transfer functions
- Technological components
- Interpersonal processes
- Releasing conditions
- Changing standards

Want to be an innovator? Want to develop an innovative culture? Get in line! Everyone wants to innovate breakthroughs that someone else has generated. Here's the price—are you:

- Smart enough to pick a generator winner to transfer from?
- Knowledgeable enough to understand the power and direction of technological components?
- Emotionally together—without ego—and interpersonally related enough to share and merge ideas with others?
- Sophisticated enough to press the release mechanisms in one's own restrictive conditions?
- Committed enough to change to become a change agent?

If you can meet innovator requirements, then you can share in its benefits. "Running in the draft" of the generators, you and your culture will have the opportunity to contribute transfers to business benefits and, in return, reap your own financial benefits.

27

The Commercializers

The "commercializers" are a culture all their own. They do not respond to the substance of the technological applications delivered to them by the innovators so much as they respond to the requirements of customer populations in the marketplace. As such, their curve is independent of the generators and innovators as they seek to "peak" and culminate the commerciality of products and services. We will conclude our illustration with the commercial evolution of the semiconductor or microchip.

Bill Gates and his company, Microsoft, are the exemplary commercializers of our time. Consciously dedicated to becoming "the IBM of software," it is said that Gates unconsciously accepted the mediocrity that came with its marketing genius. The software factory that Gates built allowed him to personally control an enormous software development organization: a system that produces operating systems, programming languages, and applications. But it came with a price: the loss of generativity and even top-flight innovation!

It is generally acknowledged in the information technology field that Microsoft is a superior marketing company, but not a leading technical innovator. According to Charles Simonyl, former head of software applications, Microsoft is not even trying to be the leading innovator. Instead, Microsoft is "trying to build core technologies that can be used in lots of products" so that "the ability to synthesize, organize, and get people to sign off on an idea or project" is most valuable.

In fairness to Gates and Microsoft, they are probably the greatest marketers and intrapreneurs in history. Customers trust that their products are slightly innovative now (see Figure 55). Moreover, the customers trust that, with their 2,000+ engineers scouring the world for innovations, Microsoft will pursue the innovative track within commercializing. As illustrated in Figure 55, scaling both GICCA and "gicca" from "1" to "5"—with "5" for generating and "1" for attenuating—we may rate Gates's products and corporation at 3.4: innovative (i) within the commercializing (C) phase. This may be the most powerful model for successful commercializing: to commercialize by being slightly innovative.

More than anything else, then, Bill Gates and his company are commercializers and marketers. In order to discharge marketing functions at the highest levels, they need to position product-lines in the marketplace. This means dedicating corporate capacities to meeting, exceeding, or creating marketplace requirements.

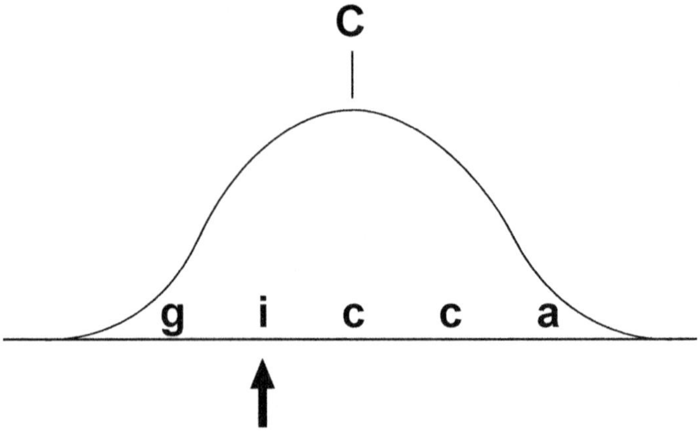

Figure 55. Innovation (i) within Commercialization (C)

In Figure 56, we have illustrated marketplace requirements in terms of "The New Capital Development" or NCD systems:

- **MC** or marketplace capital
- **OC** or organizational capital
- **HC** or human capital
- **IC** or information capital
- **mC** or mechanical capital

Of course, all of these NCD systems may be subdivided in terms of more detailed requirements.

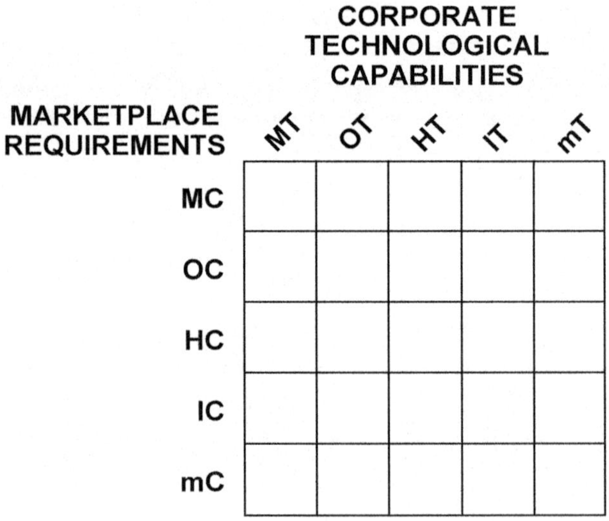

Figure 56. New Capital Development Positioning Matrix

We have also illustrated the corporate capacity to develop these NCD resources:

- **MT** or marketplace capital development technologies
- **OT** or organization capital development technologies
- **HT** or human capital development technologies
- **IT** or information capital development technologies
- **mT** or mechanical capital development technologies

Thus, the NCD capacities are dedicated to the NCD functions.

In the case of Gates's commercial empire, it is addressing information capital or IC requirements with information technology and mechanical technology capacities. It has great difficulty dealing with the IC requirements defined by information modeling. It has not even begun to address human capital or HC requirements.

Moreover, Gates's corporation must also address the phases of the market curves in which it finds itself: commercializing running into commoditizing. In order to continue to stay at the commercial center of the market, the corporation needs to develop corporate capacity in IC and HC, let alone organization capital or OC in the not-too-distant future.

The commercializers also relate to the expanding market curve (see Figure 57). Like the generators and innovators, this requires physical, emotional, and intellectual resources. However, unlike the market leaders, the commercializers seek to culminate the market curve that the leaders have initiated. In other words, the commercializers are committed to actualizing the potential of the market curve. In short, the commercializers *are* the market. Lest anyone believes that the commercializer's market mission is trivial, let them understand that this is the single most difficult market accomplishment. Those who endure, like Bill Gates, are continuously repositioning themselves to maximize meeting the requirements of customer benefits while achieving the values of profitability benefits.

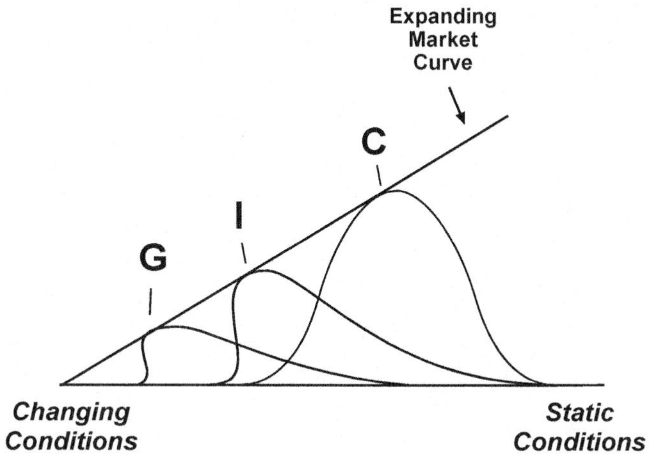

Figure 57. The Commercializer Curve

In this context, the commercializer may be defined as follows:

- Dedicated to marketing functions
- Implemented by dimensional components
- Enabled by individual processing
- Nested in selective conditions
- Measured by diversity standards

These, then, are the operations of the commercializer in both market phase and cultural or personality type:

- Marketing functions
- Dimensional components
- Individual processes
- Selective conditions
- Diversity standards

Want to be a commercializer? Want to create a commercializing culture? Everyone does these days! Pay your dues! Be...

- Smart enough to pick the transfer winner!
- Knowledgeable enough to understand the dimensionality of components!
- Persevering enough to process individually!
- Sophisticated enough to be selective in one's conditions!
- Open enough to diversity!

If you can meet commercializer requirements, then you and your culture can share in its benefits. Capitalizing upon the real work that went before you, you will receive more benefits than you make investments. But without the generators and innovators, you will have to live your lives as Gates and associates, in perpetual fear of "two guys in the garage."

28

The Generativity Team

The GICCA Team or, as we shall soon discover, "The GIC Team," is the way our American culture most effectively generates wealth. Succinctly, the keys to success lie in the interdependent and synergistic or mutual growth-producing relationships that we develop between generators, innovators, and commercializers. In this regard, "The Generativity Pull" upon innovators and commercializers is the driving engine of an expanding marketplace. Indeed, the synergistic partnership of the three cultures ensures the continuous regeneration of the marketplace. The following is a primer on the Generativity Team or "GICCA Team." The GICCA culture types emphasize different processing systems.

The attenuators emphasize S–R Conditioned Responding Systems (see Figure 58). The S–R systems are fluidly interfaced with the mechanical tooling systems that defined the Industrial Age.

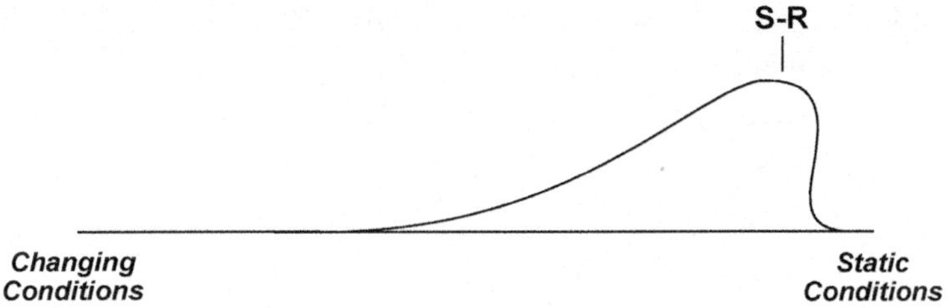

Figure 58. Attenuators and S–R Conditioning

In turn, the commoditizers emphasize S–O–R Discriminative Learning Systems (see Figure 59). Again, the S–O–R Systems are fluidly interfaced with the information branching systems that define the Information Age.

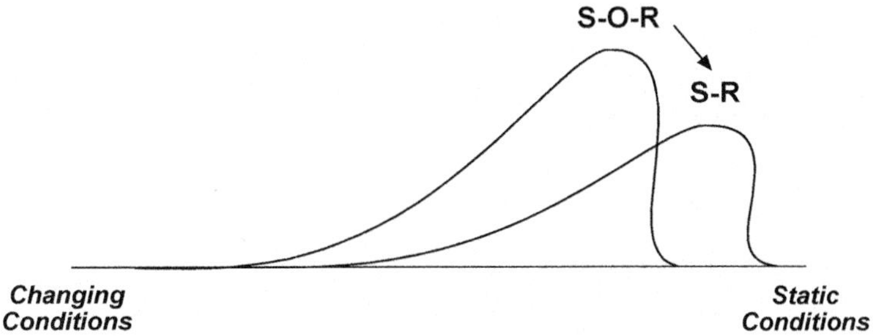

Figure 59. Commoditizers and S–O–R Learning

Similarly, the commercializers emphasize S–P–R Generative Thinking Systems (see Figure 60). In this context, the S–P–R systems are fluidly interfaced with the human processing systems that define the Ideation Age!

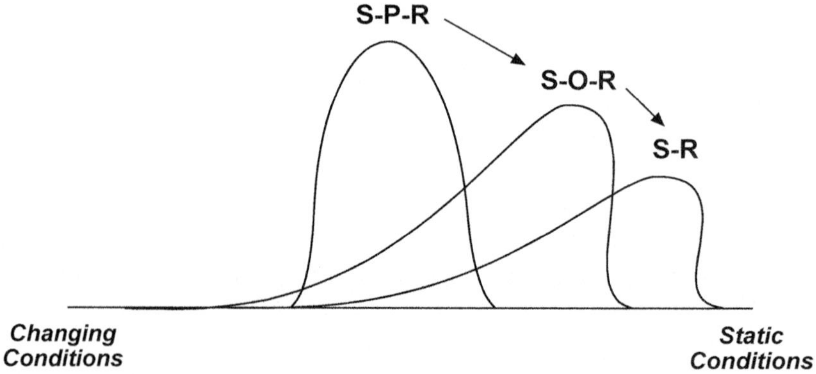

Figure 60. Commercializers and S–P–R Thinking

Likewise, the innovators emphasize S–OP–R Organizational Processing Systems (see Figure 61). In this context, the S–OP–R systems are fluidly interfaced with the organizational resource alignment systems that define the Age of Organizational Ideation.

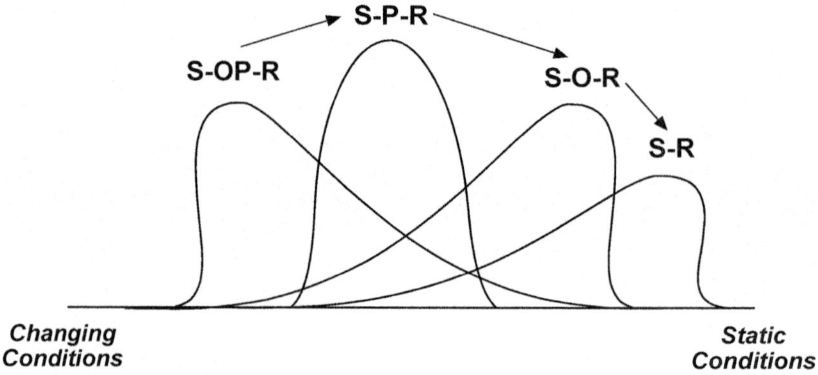

Figure 61. Innovators and S–OP–R Processing

Finally, the generators emphasize S–MP–R Marketplace Processing Systems (see Figure 62). In this context, the S–MP–R Systems are fluidly interfaced with the marketplace position systems that define the Age of Marketplace Ideation.

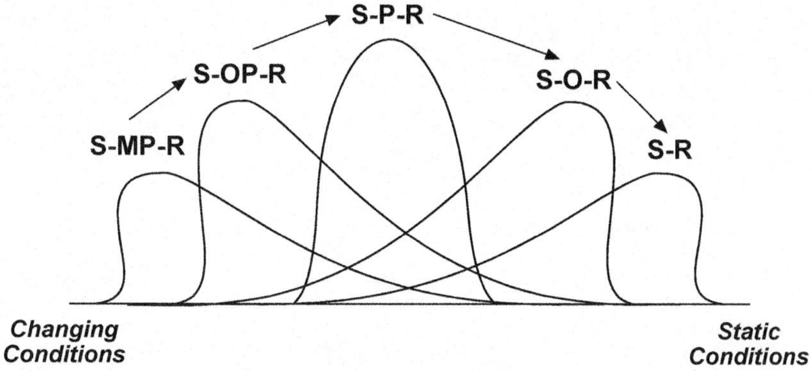

Figure 62. Generators and S–MP–R Processing

Market curves are sequenced by GICCA (see Figure 63). However, some current curves are attenuated—even truncated—by jumping in directly to one curve or another. For example, a commoditized "gimmick" may create a market curve; however, these are never powerful and never enduring. Companies that are based on these gimmicks attenuate in brief periods of time. The GICCA Sequence is enduring because each phase is empowered and released by the previous phase. Like any powerful program, if we leave out a step, people will fall in the hole: if we leave out a phase, the market will crash.

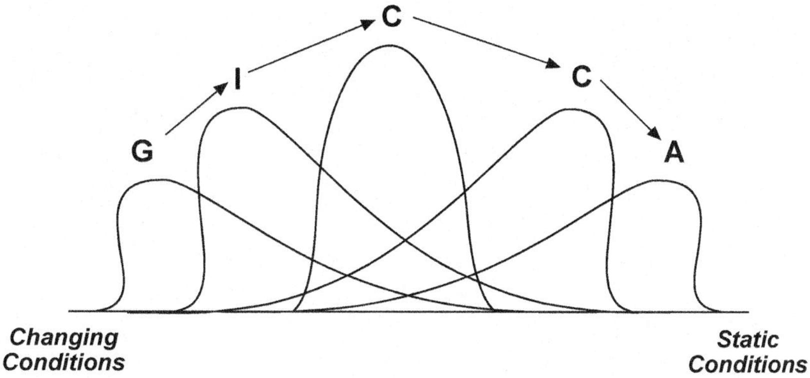

Figure 63. The GICCA Sequence

In this context, it is important to emphasize the "gaps" between market phases (see Figure 64). For example, the gap between generator and innovator is large because of the differences in their motives: the former to generate breakthroughs in substance; the latter to innovate transfers to the market. The gap between innovator and commercializer is still greater: while the innovator focuses on transfer, the commercializer focuses on customer needs. Some of the debates approach violence as each phase takes ownership of "the child" from the previous phase. Bridging these gaps is the greatest challenge to any market movement.

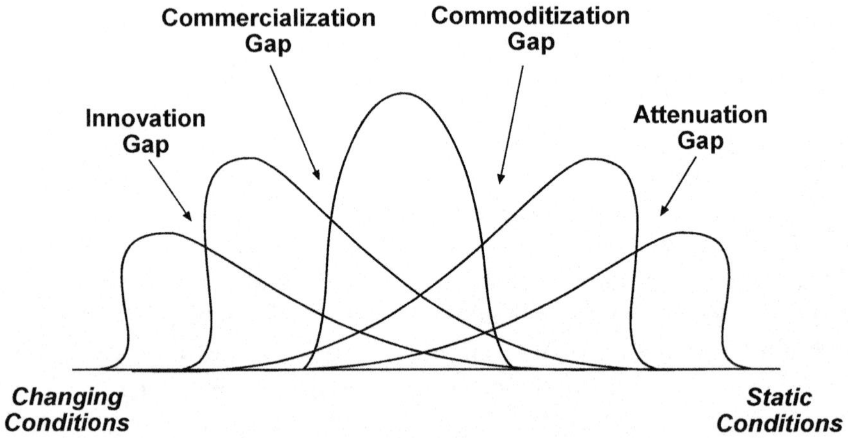

Figure 64. Bridging Gaps in Curves

The market curves are accelerating (see Figure 65). This is primarily because of intensified innovation and commercialization. In other words, this is because of the emphasis on "D" in the old "R&D": innovative development of products for commercialization. As the "innovation–commercialization" cycle intensifies, there is diminishing space for commoditization and attenuation. Indeed, as the commercialization phase declines, the commoditization is immediately set upon by attenuation that, in turn, is beset by truncation or the end of the curve.

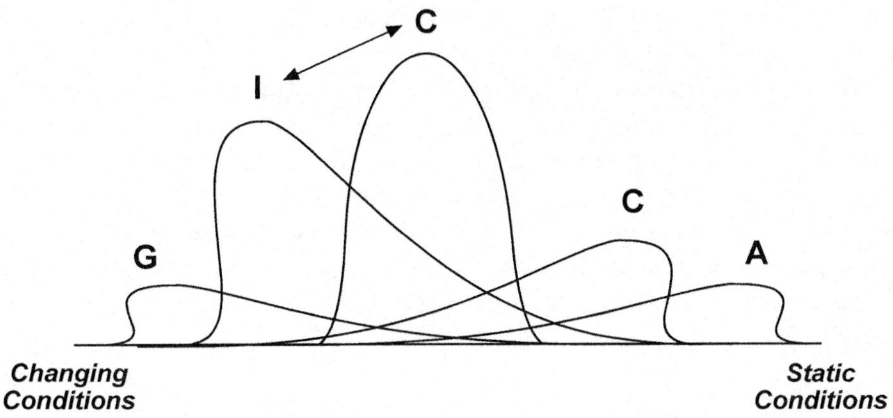

Figure 65. Accelerating Curves

However, such "Innovator–Commercializer Curves" lose the original source of their vitality (see Figure 66). Increasingly, generativity becomes the market differentiator. With the de-emphasis on "R" in the great old "R&D" houses, the market lost its differentiators. Consequently, "Innovator–Commercializer Cycles" are of shorter and shorter durations. Where innovators could once hold a competitive edge for years, they are now fortunate to have a comparative advantage for months. Generators become the great differentiators just as it was during the early part of the 20th century with Edison and Ford and during mid-century with Bell Labs and the Watson and Xerox Centers.

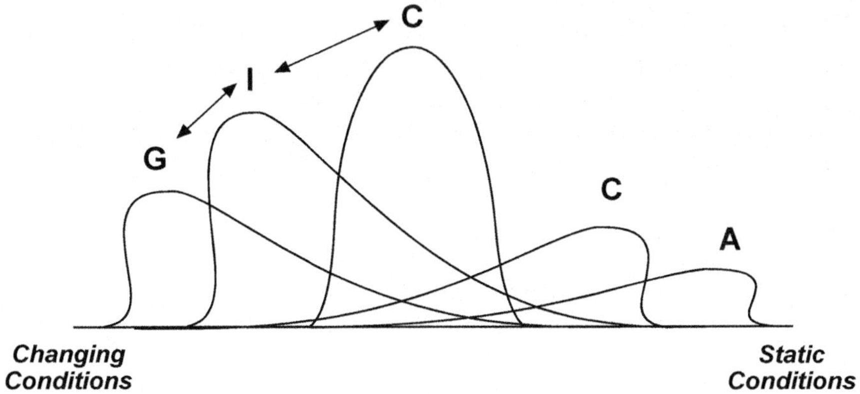

Figure 66. Intensifying Generativity

Generativity requires a price far greater than most populations are willing to pay (see Figure 67). Above all else, generators commit to become "one and growing" with their substance. Just as we are presenting to you the operations of the GICCA curve, so must generators define the operations of their substance: functions, components, processes, conditions, standards. In so doing, the generators exert a "pull" upon the innovators and commercializers. The innovators must now discharge their transfer functions technologically, and the commercializers must now define their marketing functions operationally. All move toward the generative source of the curve.

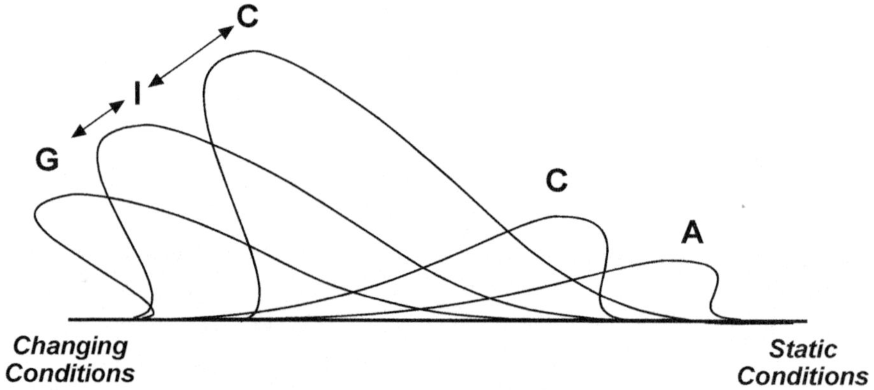

Figure 67. The Generativity Pull

All curves move interdependently and collaboratively toward change conditions (see Figure 68). In particular, the generators, innovators, and commercializers relate synergistically: each contributes to the other's growth. We label this "The GIC Team": it overcomes all gaps between curves by Continuous Interdependent Processing or CIP. We label this "The CIP Vision." Together, "The GIC Team" and "The CIP Vision" contribute synergistically to continuous change in market conditions. Ultimately, we all have to choose:

- between becoming dependent minions of periodic eruptions of change because of our myopic vision and conditioned responding to that vision; or
- becoming partners in generating, innovating, and commercializing continuous change because of our continuous interdependent processing!

The choice is ours! We hold our continuously changing corporate destinies in our minds!

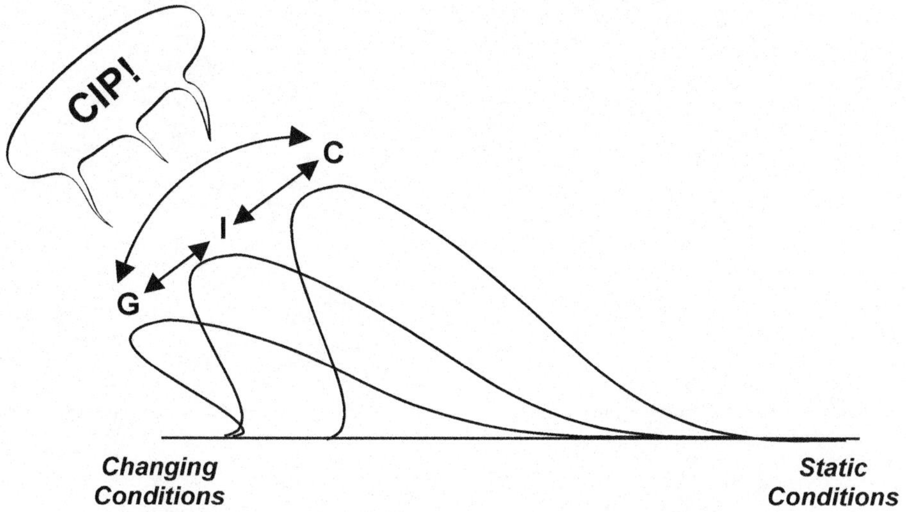

Figure 68. Continuous Interdependent Processing

29

NCD-Driven Economics

Behind the GDPs of nations and the profitability of corporations lie the sources of economic productivity and growth: growth in profitability, growth in market share, growth in market size. Our X-ray vision reveals these sources of performance and growth as the ingredients of new capital and their development and integration.

Having been defined as "most important," the historic power of financial capital has devolved to account for about 15% of **economic productivity growth** or **EPG**. It is the other 85% of EPG that we are addressing in **new capital development** or **NCD: MCD, OCD, HCD, ICD, mCD.**

Throughout history, individuals and companies have contributed NCD ingredients to economic enterprises. However non-systematic and neutralized in their potency, these nascent forms of NCD have made all the difference in the successes of the enterprises.

Now, with the comparative contributions of IT reduced to "utility computing," it becomes imperative to view these NCD dimensions in sharpest relief. The ICD-driven and HCD-driven operations are now poised to supplant the IT-driven and mCD-driven economic operations as prepotent sources of comparative advantage.

We may view the development and integration of NCD in the evolution of economic operations from mCD through ICD to HCD, OCD, and MCD. We will present these images in prototypes:

- **mCD**-driven economic operations
- **ICD**-driven economic operations
- **HCD**-driven economic operations
- **OCD**-driven economic operations
- **MCD**-driven economic operations

mCD-Driven Economics

The first set of these prototypical images are the **mCD**-driven economic operations typical of the early Industrial Age (see Figure 69). As may be noted, the mCD operations may be defined as follows:

> **Delivery task functions are discharged by production-driven components enabled by outputting processes.**

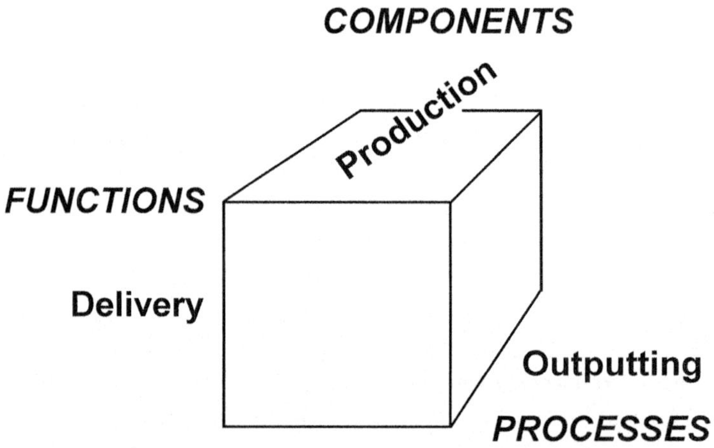

Figure 69. mCD-Driven Economic Operations

Another way of defining these mCD operations is that they are dominated by S–R Conditioned Responding Systems: like primitive assembly lines, everyone involved makes conditioned responses to the stimulus conditions they receive; in turn, everyone hands off their products as stimuli to the next recipients. All participants respond reflexively. There is no expectancy for participants to process intellectually. They are not empowered to do so.

ICD-Driven Economics

The second set of these prototypical images are the ICD-driven economic operations typical of the Information Age (see Figure 70). As may be noted, these **ICD** operations may be defined as follows:

> **Supervisory objective functions are discharged by technology-driven components enabled by planning processes.**

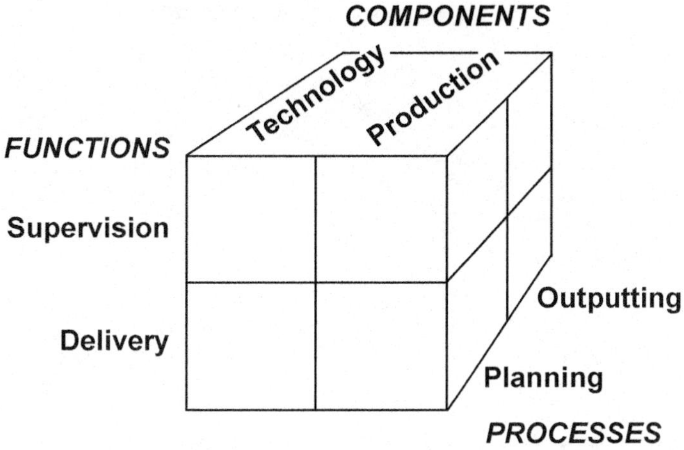

Figure 70. ICD-Driven Economic Operations

Another way of defining these ICD operations is that they are dominated by S–O–R Discriminative Learning Systems. They operate in branching systems with supervisory objectives empowered by technology and enabled by plans. All participants respond discriminatively. The participants process with repertoires of conditioned responses from which they emit those appropriate to the stimuli. They lack a powerful vision of ICD with which to do more.

HCD-Driven Economics

The third set of these prototypical images are the HCD-driven economic operations that characterize the Transitional Age in which we now find ourselves (see Figure 71). As may be noted, these HCD operations may be defined as follows:

> **Management systems functions are discharged by resource-driven integration components enabled by processing systems.**

Another way of defining these HCD operations is that they are powered by S–P–R generative processing systems. They operate as "human brainpower": continuous processing to expand possible alternatives and narrow to preferred courses of action.

The management function serves to empower people to design systems to integrate resources and achieve goals. All participants process to generate the ICD that makes the discriminative learning systems work efficiently and effectively.

This is, indeed, a Transitional Age, lying between the Ages of Information and Ideation. However, here we are caught in a double-bind condition: we do not have the ICD to empower HCD; we do not have the HCD to generate ICD.

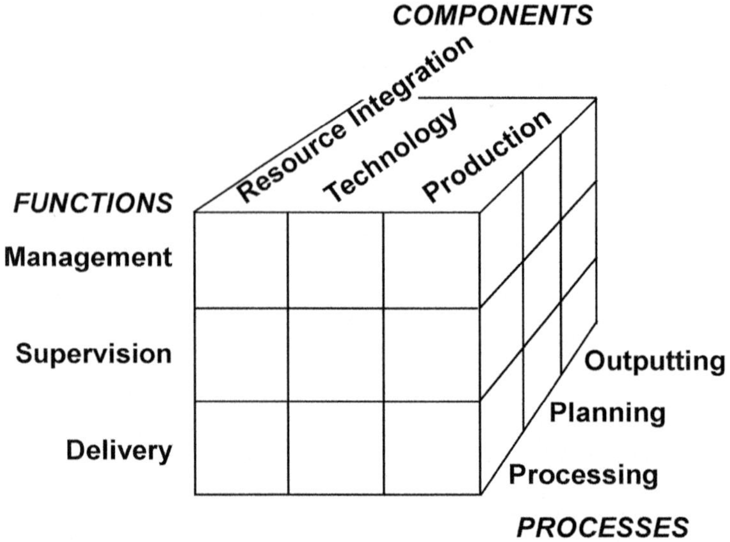

Figure 71. HCD-Driven Economic Operations

In the Information Age, we were able to deliver the right data to the right people at the right time. The problem is that we did not understand the meaning and rigor of ICD operations.

Now in the Age of Ideation, we are able to define the components and functions of information that we need. The problem is that we have little idea about the meaning and rigor of HCD processing operations.

While there are many false prophets of HCD, none but one will guide us. There is only one ingredient that makes humans truly "capital," and that is our generative processing skills: the power to create better ideas!

OCD-Driven Economics

The fourth set of these prototypical images are the **OCD**-driven economic operations that also characterize our Transitional Age (see Figure 72). As may be noted, these OCD operations may be defined as follows:

> **Executive architectural functions are discharged by marketing-driven components enabled by inputting systems.**

Another way of defining these OCD operations is that they are aligned by S–OP–R Organizational Processing Systems, a definable set of organizational operations: higher-level operations **nesting** and **encoding** lower-level operations; all operations **rotating** appropriately to become functions commanding the continuous realignment of dimensions of other resources.

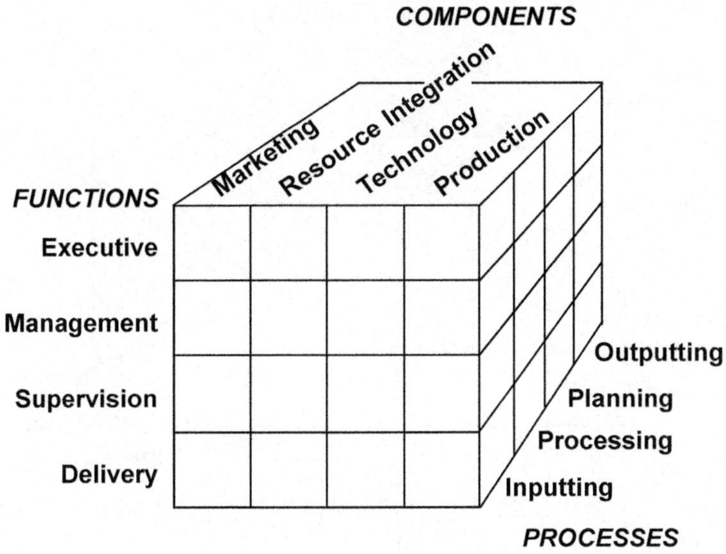

Figure 72. OCD-Driven Economic Operations

The executive function serves to design the organizational architecture to guide the systems to discharge their changing functions. All contributors process generatively to bring all NCD systems into the synergistic relations that make each system contribute maximally: MCD, OCD, HCD, ICD, mCD.

Again, we are now caught in a triple-bind condition: we not only lack systematic and comprehensive ICD and HCD, we also lack OCD alignment systems.

Once again, while there are many premature philosophers, none will guide us to actualize the contributions of OCD. There is only one ingredient that makes organizations truly capital: the ability to continuously realign resources to achieve better ways of accomplishing its changing functions.

MCD-Driven Economics

The fifth of these prototypical images are the **MCD**-driven economic operations that culminate our Transitional Age (see Figure 73). As may be noted, these MCD operations may be defined as follows:

> **Policy positioning functions are discharged by leadership-driven components enabled by goaling systems.**

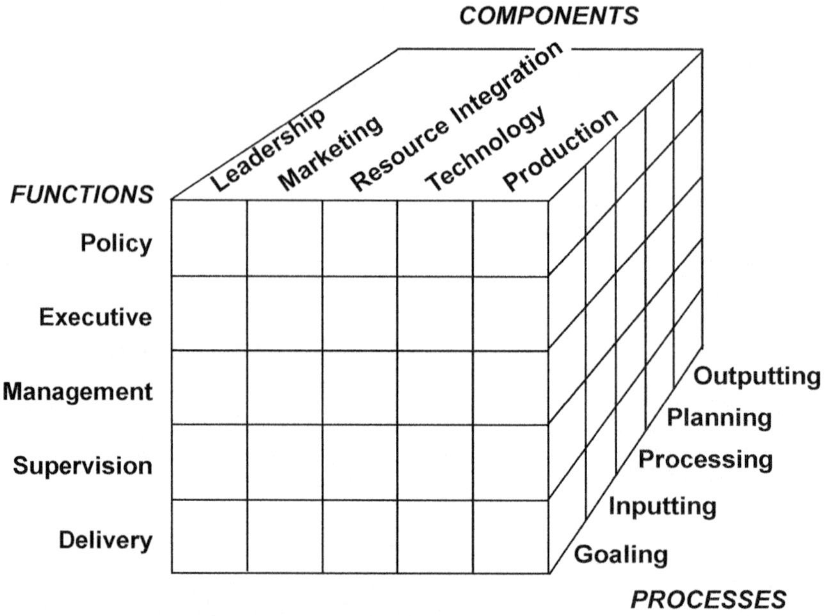

Figure 73. MCD-Driven Economic Operations

Another way of defining these MCD operations is to say that they are positioned by S–MP–R Marketplace Processing Systems, a definable set of interactive marketplace operations: continuously evolving customer preferences dictating constantly changing marketplace requirements and continuously evolving organizational values and capacities.

The policy function serves to position the organization to maximize meeting both consumer requirements and organizational values at the highest levels. All leaders process generatively to refine and relate all continuously changing, potentially synergistic operations.

Processing-Driven Economics

Summatively, our X-ray vision reveals that all economic operations are processing operations (see Figure 74):

- **S–MP–R Marketplace Processing Systems**
- **S–OP–R Organizational Processing Systems**
- **S–P–R Generative Processing Systems**
- **S–O–R Discriminative Learning Systems**
- **S–R Conditioned Responding Systems**

These systems are iterative and interactive. They apply to all functions, components, and processes.

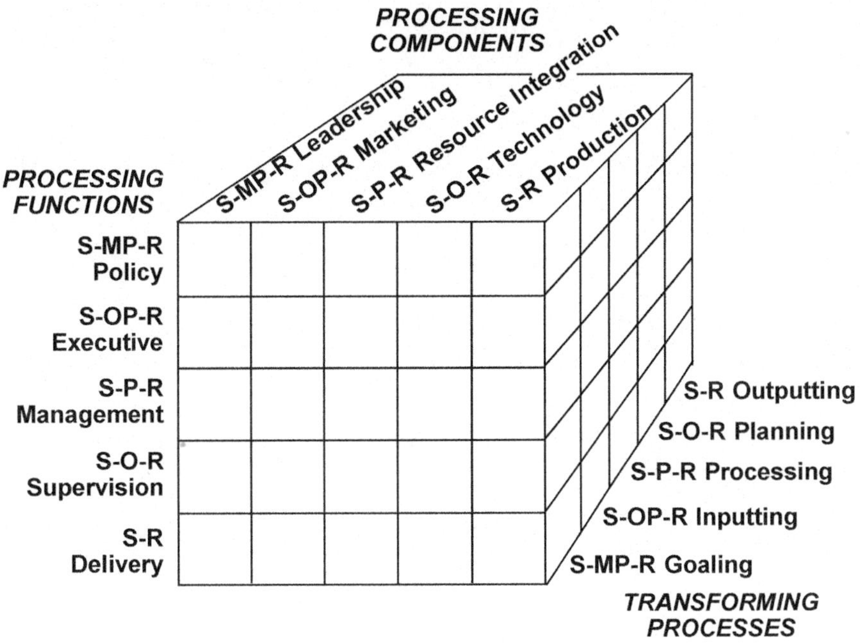

Figure 74. Processing-Driven Economic Operations

In their simplest forms, all processing systems are "nested" and, thus, "encoded" by the marketplace positioning of S–MP–R. This means that all processing systems can be aligned and rotated to implement marketplace positioning.

30

NCD-Driven Ages

Another way of summarizing these NCD-driven economic operations is to use a linear table of Economic Ages (see Table 7). As may be viewed, all economic processing operations are dedicated to NCD: MCD, OCD, HCD, ICD, mCD.

Table 7. Drivers and Processing Systems of Different Economic Ages

Economic Age	NCD Driver	Processing System
Industrial Age	**mCD**	**S–R Conditioning**
Information Age	**ICD**	**S–O–R Discriminating**
Transitional I	**HCD**	**S–P–R Generating**
Transitional II	**OCD**	**S–OP–R Aligning**
Transitional III	**MCD**	**S–MP–R Positioning**
Age of Ideation	**MCD** **OCD** **HCD** **ICD** **mCD**	**S–MP–R** **S–OP–R** **S–P–R** **S–O–R** **S–R**

All NCD systems are directed by MCD or marketplace positioning, and are in turn aligned and rotated to develop all other forms of new capital: OCD, HCD, ICD, mCD.

Probabilities Differences

Of course, few systems fall nicely into these prototypes. Nevertheless, all successful economic systems will develop and integrate the NCD processing systems. They become possibilities economic systems when they address the marketplace with all processing responses in their operational repertoires.

One helpful way to conceive of the NCD processing systems is by rating them. Utilizing five-level scales, we can conclude the following probabilistic differences:

- **Level 1** = 1 cell
- **Level 2** = 8 cells
- **Level 3** = 27 cells

- **Level 4** = 64 cells
- **Level 5** = 125 cells

From this alone, we may conclude that operations like **mCD** and **IT** systems are severely restrained.

Possibilities Differences

However, when we reason deductively from our comprehensive **NCD** model, we may approximate **possibilistic differences.** With five sources of effect, each system accounts for 125 cells. In total, **NCD** defines its universe of cells as 125 to the 5th power, or 125^5 (see Table 8).

Table 8. Levels of NCD-Driven NCD Systems and Universes of Cells

Levels of NCD-Driven Systems	Multipliers	Universes
mCD	125	125
ICD	125^2	15,625
HCD	125^3	1,953,125
OCD	125^4	244,140,625
MCD	125^5	30,517,578,125

In quantitative terms, this means that comprehensively developed and integrated NCD systems would be billions of times more powerful than minimally developed systems such as **mCD** and **IT.**

Qualitative processing is still another story! With processing embedded and interactive with, between, and among all cells, the power might approach the ultimate comparison: **Infinity > Infinitesimal.**

Processing Systems

On the face of it, the possibilistic differences appear overwhelming—indeed, impractical to many business people. In practice, the processing systems render the objectives operational (see Table 9):

- **MCD** generates the corporate positioning in which all **NCD** is **"nested"** and by which all **NCD** is **"encoded."**
- **OCD** generates the organizational alignment within which all **NCD** is **"aligned"** and then **"rotated."**
- **HCD** generates the human processing that **"expands"** and then **"narrows"** new images of operations.
- **ICD** generates the information modeling that **"dimensionalizes"** and then **"operationalizes"** productive images of operations.
- **mCD** generates the mechanical tools that **"technologize"** and, then, **"produce"** the products of the operations.

As illustrated, the operations can be rotated to individual or team work stations and prioritized by functions.

Table 9. Processing Systems and Applications of NCD Goals

NCD Operations	Processing Systems	Applications
MCD	S–MP–R	Corporate Positioning
OCD	S–OP–R	Organizational Alignment
HCD	S–P–R	Human Processing
ICD	S–O–R	Information Modeling
mCD	S–R	Mechanical Tooling

Processing Outcomes

At a minimum, we may conclude that approximations of comprehensively developed and integrated NCD models are exponentially more powerful than the primitive mCD and IT-driven operations. Whether in business or with nations, we may consider the implications of limited-cell cultures relating dependently and reactively—even violently—to advanced, interdependently integrated, generative processing cultures!

Relatedly, it is not the terrorist power of limited-cell societies that we fear. It is the lack of fulfillment of our own human mission:

- Like NCD, to actualize the ultimate processing power of the many billions of neurons in human brainpower
- Like the cells of NCD, to actualize the simultaneous processing with many thousands of other neurons in our brainpower

In transition, few organizations have approached the robust power of fully developed and integrated NCD. Perhaps some people at Airbus have had glimpses of it. Keep on processing!

No nations have even entertained the power of interdependency in the form of processing-driven NCD operations, even as the means of accomplishing free enterprise economics. Moreover, the processing vision has been obfuscated by most multinational corporations that, having achieved success with IT-driven operations, seek to utilize their limited S–O–R systems to restrict the operations of others in the marketplace. The vision is further clouded by the earnings in financial capital of the primitive operations of resource-based companies and nations, such as the oil-producing nations now engaging in "the last dinner of the dinosaurs!"

The NCD systems are prepotent! Pursued intensively and extensively, they will further reduce the economic growth contributions of financial capital to mere catalytic levels.

In so doing, NCD will dwarf the contributions of the mechanical and relational databased driven IT operations of the Age of Information.

Instead, NCD will generate the processing-driven operations of the Age of Ideation.

Our ideational products will be ICD:

- Information models that constitute the basis of all science and technology and, thus, economics
- Information models that are created continuously by the interdependently and synergistically related HCD, OCD, and MCD generating systems

This will introduce us to "the infinity reaches of capital wealth!"

Welcome to the possibilities economy!

Part VIII

Generativity Positioning

31

Generativity Positioning Systems

The generativity economy begins with positioning. Without positioning, the organization or economy cannot gain or maintain comparative advantage over others in the marketplace. Another way of saying this is that positioning differentiates our economy's unique contributions.

However, in today's world, we cannot talk about positioning as simply differentiating ourselves from our competitors. Competition is the function of static resources and independent attitudes. Independence and competition, as we once practiced it, will not work any more.

Today, we position as interdependent partnerships with nearly everyone: customers, customers of customers, vendors, suppliers, and, yes, even entities that we might still consider "competitors." Collaboration is the function of interdependence. Today, positioning differentiates our interdependent collaborative contributions.

All business policymakers will concede that the success of a corporation begins with its positioning in the marketplace. Yet all of these top executives will also admit that they were never taught "positioning." Many employ "legacy positioning" by "sticking to the knitting" and adding new related product and service lines. Others employ "existential positioning" by "cleaning the slate" and entering those markets with highly leveraged opportunities. There are many other strategies in between these extremes—some very personal and idiosyncratic. But none are systematic in positioning their corporations in the marketplace!

One of the great ironies of business is that we know so much about the technologies we employ to produce our product and service lines, but so little about the technologies of the business practices that determine the success or failure of our businesses. That is partly the reason why our business successes are random and seldom replicated.

Increasingly, top executives are driven by fast-changing marketplace requirements. As breakthroughs evolve through the rapid market life cycle, executives must address the derivative requirements of these breakthroughs. Marketplace requirements are not simply based on corporate values or executive choices. The marketplace itself "speaks" and its requirements must be understood. With this understanding, corporations take an index of their capacities to address these requirements. Corporations need technological capabilities in precisely the same areas as the requirements.

To enable the corporation to mobilize its technological capacities to address its marketplace requirements, executives must understand the processes of marketplace positioning. All positioning to enhance corporate capacities to meet marketplace requirements is accompanied by implications for placement in the life cycle of the market.

We must remember: no organization is an independent entity. Our organizations are inextricably linked in generating the "New Market Life Cycle," from generativity and innovation through commercialization to commoditization and attenuation. Marketplace positioning must relate customer requirements with producer technologies, all within the context of the New Market Life Cycle.

Three points are particularly important here:

- **First, understanding the requirements of the marketplace is crucial to positioning.** In effect, marketplace requirements are the values of the marketplace. This is analogous to career placement: we have our values, and the job has its own values; the values of the job become the requirements of the person. In marketplace positioning, the marketplace values become the organization's requirements. Our external mission is to position our organizations to meet or to exceed these marketplace requirements.
- **Second, gaining the technological capability to meet marketplace requirements is our internal mission.** This, too, is analogous to career placement: if we have the right skills, knowledge, and attitudes, we can get the job. Mobilizing technological capacity to meet marketplace requirements is the essential function of the organization.
- **Third, how we address the requirements of the marketplace with our organizational capacities has implications for our placement in the market life cycle.** Beginning with generativity, possibilities are transformed into marketplace applications by innovation, and culminate in profitable products and services by commercialization. An interdependent partnering relationship among these three phases of the market life cycle is essential for powerful positioning in the marketplace.

Technological capabilities and marketplace requirements are both "moving targets" in the New Market Life Cycle. Whatever economy *best understands the market's requirements, builds its technological capabilities, and partners across the market life cycle* is taking the first step toward becoming a "generativity organization."

Armed with the ideational image products of our generative processing, we may now position our organizations or economies in the marketplace. Our most powerful positioning is to be able to do something no other entity can do.

Generative–Innovative positioning was the historic positioning of the American economy in the latter part of the 20th century. Today, the same positioning requires that we offer "virtual off-the-shelf products" that customers may choose to have tailored to their needs.

The prepotent breakthrough model is the Generativity Positioning System™ or GPS™ (see Figure 75). As may be noted, GPS™ represents and implements generative–innovator positioning in the marketplace at the highest levels as follows:

Generative positioning functions are accomplished by policy-making components empowered by leadership processes.

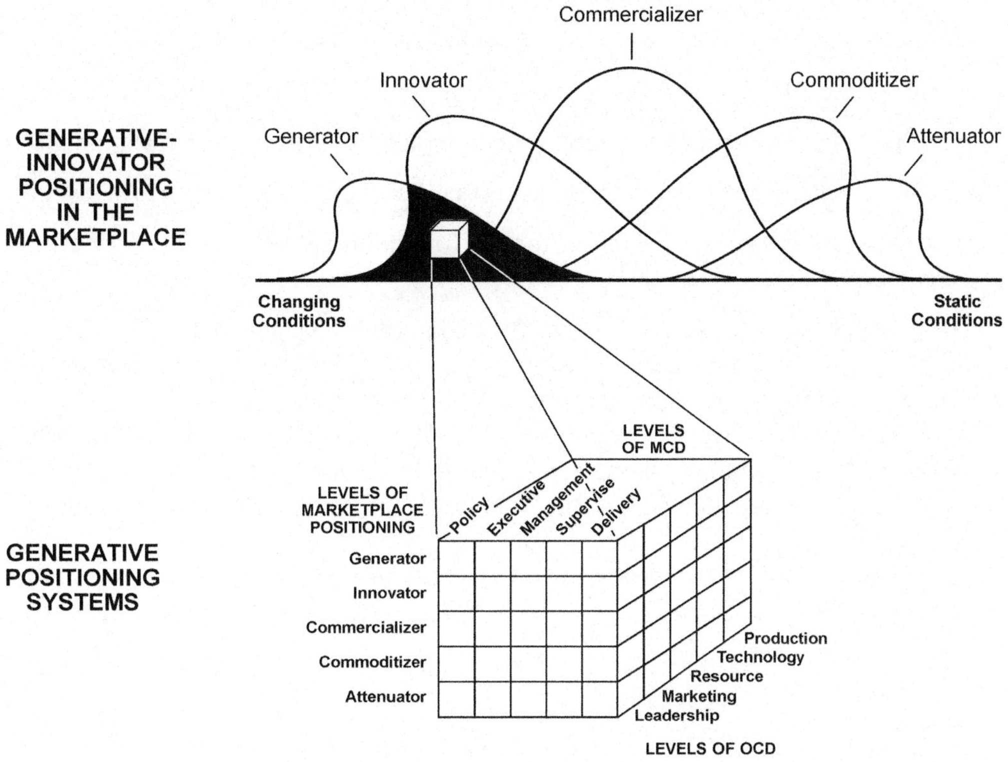

Figure 75. The Generativity Positioning System™ Representation in the Marketplace

The most critical function of this GPS™ source cell is implementing the Requirements–Capacities Matrix: the capacities of the organization are compared against the requirements for generative–innovative positioning in the marketplace. Clearly, generativity positioning is going to require the highest levels of new capital development to establish and continuously maintain positioning.

In GPS™, the customer may select from potentially an infinite array, those products that are (1) most closely tailored to meet the customer's values, and (2) for which the customers are empowered by tailored capacities to meet the product's requirements for managing.

For example, the customer may view electronic representations of all kinds of products—homes, cars, equipment—via Computer-Assisted Design. The customers may then select and "tailor" those products to meet their needs.

The generative marketplace of the future will be driven by the Generativity Solution—the generativity of breakthrough initiatives and architectural designs (see Figure 76). Anything of value that can be conceived can be achieved in Generativity Positioning Systems™ of new capital development.

- **MCD** or continuous marketplace position for comparative advantage
- **OCD** or continuous organization alignment with marketplace positioning
- **HCD** or generative human processing to implement organizational alignment
- **ICD** or schematic information modeling to reflect generative human processing
- **mCD** or operational mechanical tooling to implement schematic information modeling

The actual design and assembly of the products will be "farmed out" to "MOHIMs" for the New Capital Development Systems just as OEMs or original equipment manufacturers currently organize their production efforts.

The paradigm shift to GPS™ requires that all NCD Components contribute to the substance of the products being produced and delivered. The criteria of measurement revolve around increasing the "size of the pie" rather than redistributing the "sizes of the pieces."

In this context, we will have once again placed Americans back in the leadership role:

- The American learners with expansive acquisition of operational knowledge and skills
- The American workers with elevating levels of applications and transfers in producing products and delivering services
- The American entrepreneurs with escalating levels of transfer initiatives and risk in generating wealth
- The American citizens with accelerating levels of enlightenment and participation in governance and politics
- The American leaders with potentially infinite levels of interdependent generativity in all areas of policy making

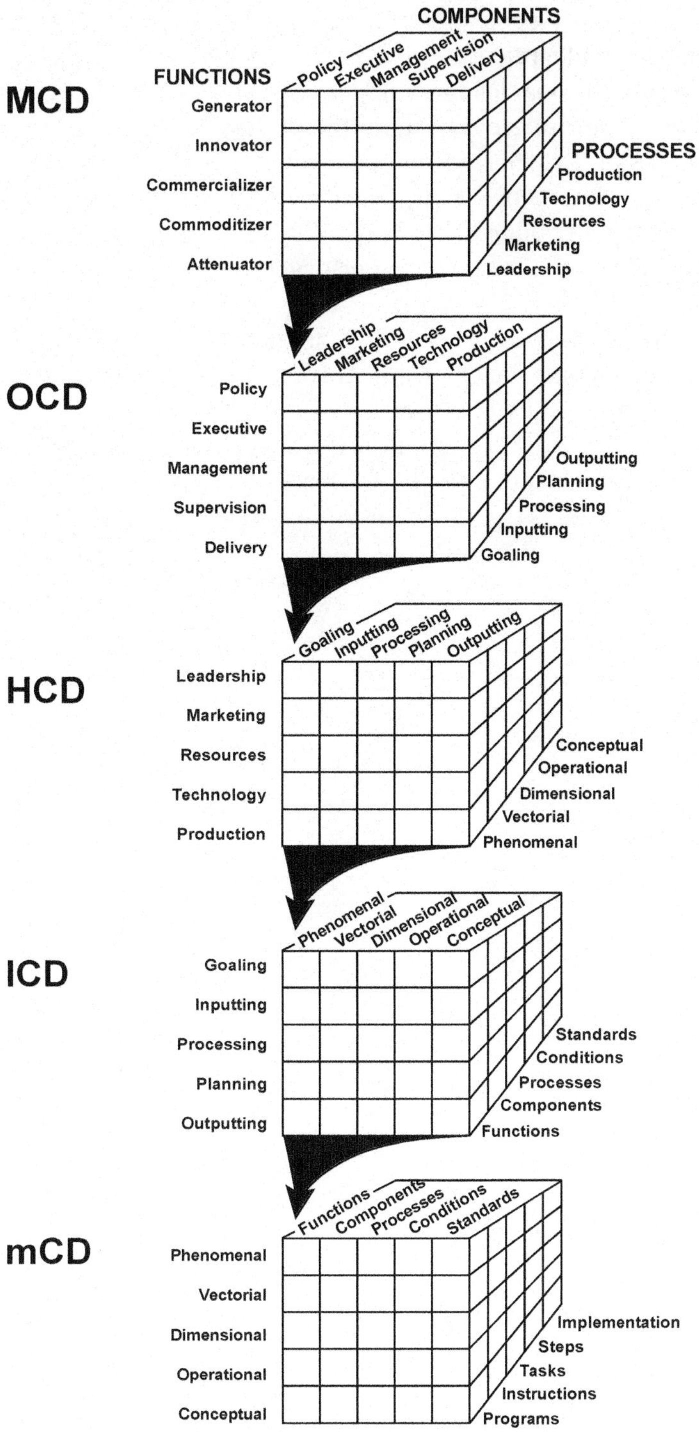

Figure 76. New Capital Development Systems

In other terms, we will have a fulfillment of human generativity, reinvigoration of the new capitalism, and a renaissance of civilization.

Just as the organizational leaders on Main Street and Wall Street need to learn the ingredients of corporate growth, so do the national leaders in America need to learn the ingredients of "Real Socioeconomic Growth."

Just as the underclass people marginalized by "the cycle of social failure" need to learn how to become productive individually, so do those marginalized by "the cycle of financial failure" need to learn to perform productively and profitably in our times.

Rather than to pit Main Street versus Wall Street versus Constitution Avenue, we must all learn to become generative in entrepreneurial ways where each grows so that all have the opportunity to grow.

In transition, the generativity solution empowers us to solve all human crises. The generativity solution is the "human solution."

32

The New Economic Paradigm

More than 30 years ago, the embattled Lee Iacocca, head of Chrysler, humbly submitted his application for a bridge-loan of $1.2 billion. While the idea of the federal sector supporting the private sector was controversial, Mr. Iacocca and Chrysler responsibly paid back the loan with interest to the taxpayers of America.

Today, all elements of society—public and private sectors alike—are competing at the trough for their "fair share" of the Troubled Asset Relief Program (TARP) and other gimmicks that have been dreamed up by our "faux leaders" who would not have us pay for their friends' transgressions in the "self-adjusting and self-regulating markets." They would have our children and grandchildren pay a tab already expanded by the U.S. Treasury to many trillions of dollars and counting. Yes, thousands of times the amount asked for by Mr. Iacocca!

So, what do we get for these investments of our progeny? The best that I can figure out follows:

1. Economic Initiatives

We are going to make trillions of dollars of investments in infrastructure rebuilding and environmental rehabilitation. These, in turn, are going to produce "make-work" jobs in order to reinvigorate the economy and hire the unemployed. Notwithstanding the "IQ lid" that Industrial Age jobs place on this generation, nor the dollar burden that the cost imposes on future generations, there is no evidence that this socialistic strategy will work. Massive investments of this kind failed during the Great Depression of the 1930s and, indeed, exacerbated conditions that precipitated the crash of 1937, the low-point of the Depression. The spending-spree never did bring us out of the Depression. It was not until the anticipatory war years of 1939 and 1940 that the economy began to recover. Moreover, regarding the financial markets, it took 25 years and a World War for the Dow Jones Averages to return once again to the heights they recorded preceding the 1929 crash. Perhaps the greatest threat to both economic and employment growth is the loss of confidence in free trade and open markets in this modern era of globalization. Both trade and capital flows—the engines of global integration—are in reverse gear. The "depressor variable" of U.S. trade deficits has exposed the conflict between national interest and globalized business. Succinctly, the Chinese–U.S. trading relationship is the linchpin of globalization: China produces with our raw materials and our sophisticated technologies; U.S. consumes inferior products in a debtor relationship; China lends the proceeds back to the U.S. so that it can continue to consume. Nowhere do our superior technologies and production systems play a role in these trading practices.

2. Cultural Initiatives

We are going to make these massive investments in order to restore the character and confidence of the American people. Nothing could be further from the truth! The informal conferences of the 1980s and 1990s, cynically dubbed "Bretton Woods II," have already changed the character of the American people and their economics from producer to consumer. The net of this contract is the following: America plays the "parasitic host" of debtor in international trade while creditor nations purchase our debtor bonds. The effects of this consumer conditioning have already reduced our citizens' performance on those cultural indices leading to wealth generation: from collaborative cultural relating to dependency; from enlightened participative governance to authoritarian socialism; from entrepreneurial free enterprise to command-and-control protectionistic economics. Let it be known to one and all, that while we were on the verge of the highest levels of these Freedom Functions, we have been reduced to utter despair by the "Totalitarian Socialist Administration for Multinational Corporations" installed by the Bush administration in the 21st century. A continuation of this socialist tradition ostensibly dedicated to the middle class bodes poorly for the citizens of the world as well as America.

3. Community Initiatives

Perhaps the greatest source of initiative comes from the human generativity that is the source of all other new capital development initiatives. I have already visited with the leaders from Route 128 in Boston, the so-called "Silicon Valley East" of the marketplace. I inquired of them: "After information technology—what?" To a person, they answered, "Teach your children Mandarin!" And why not? They have already seen the "Economic Engines of the West" stumbling with lack of support. For over 30 years, the "entrepreneurial capital" of the world has thrived in Silicon Valley, California, with the "Generation–Innovation–Venture Capital–IPO Cycle" generating new wealth, commercializing innovation, and creating new companies and industries while spinning off millions of new jobs. This "cycle of generativity" is self-perpetuating: Entrepreneurs generate new ideas, form teams, architect and plan, and promote the ideas to venture capitalists. In turn, VCs invest, start-ups perform, and IPOs are offered. The cycle begins again with the profitability and reinvestment of the companies. The success of this driving force in our economy is over, under assault by legislators and regulators who do not understand the fragile nature of generativity. In the name of "fairness," the SEC and Financial Accounting Standards Board have put individual generativity at risk. In 2008, a miniscule number of companies from Silicon Valley are now going public, compared with an average of hundreds per year over the previous two decades.

The Generativity Package

Political attempts by super-organizations such as Asana to create "consensus science and technologies" notwithstanding, little is left from the products of the binary code systems to build upon. At GenStar, LLC, we have continued on our R&D to build human-centric information science and technologies.

After my encounter with business leaders of Silicon Valley East, my colleagues and I set about to architect generativity solutions at every level of community and cultural generativity:

- Every infant gets stimulated!
- Every child learns to discriminate!
- Every adult learns to generate!
- Every citizen becomes enlightened!
- Every worker becomes entrepreneurial!

Together, these define the community and cultural generativity standard that constitute the foundation for economic generativity.

Simultaneously, in our research and development center, we architected and demonstrated generativity solutions at every level of the generativity paradigm:

- Individual generativity systems
- Organizational generativity systems
- Community generativity systems
- Cultural generativity systems
- Economic generativity systems

Together, these generativity solutions generate the productive ingredients for spiraling socioeconomic growth.

Together, these generativity solutions constitute "The Package That Keeps on Giving!"

Part IX

Entrepreneurial Generativity

33

The Economic Freedoms

The intent of the researchers at the Heritage Foundation is to measure the degrees of economic freedoms and relate them to the levels of economic prosperity.

The economic freedoms are defined as follows:

> **"...all liberties and rights of production, distribution, or consumption of goods and services."**

A sample of the corollaries of economic freedoms follow:

- Absolute rights of property ownership and constitutional law
- Fully realized movements for labor, capital, and goods
- Absolute absences of coercions or constraints of economic liberty

In short, individual entrepreneurs would be free to produce and consume in any way under the rule of law. Moreover, their freedom would be at once protected and respected by the state.

While the necessity for governmental intervention for the "public good" is stipulated, governmental coercion that rises above the minimal level is considered "corrosive to freedom"—most notably economic freedom.

The 10 specific economic freedoms measured in the 2009 Index of Economic Freedom are identified below. Each of the freedoms is individually scored. A country's overall economic freedom score is an average of its scores on the 10 individual freedoms:

- Business freedom
- Trade freedom
- Fiscal freedom
- Government size
- Monetary freedom
- Investment freedom
- Financial freedom
- Property rights
- Corruption freedom
- Labor freedom

In perspective, the 10 factors emphasize governmental activities that facilitate or retard free economic enterprise. They are not, themselves, indices of the governance processes for generating initiatives that enhance or constrain free enterprise. To obtain these indices, we must study the nature of governance processes, ranging from totalitarian to free democratic. Nor are these 10 factors indices of economic performance. To obtain performance indices, we must study the relationships of these broad factors with individual, or per capita GDP.

The sources of economic growth and prosperity may be summarized globally with data on economic freedom and per capita GDP (see Figure 77). **The correlation relating "Economic Freedom Scores" to GDP indices is +0.69. This means that we can account for approximately 50% of the variance in economic prosperity and growth by a country's level of free enterprise:**

- **Those countries that engage in free enterprise are economically prosperous and growthful.**
- **Those countries that emphasize command-and-control economies are economically poor and stagnant.**

In summary, Economic Prosperity is highly related to Economic Freedoms.

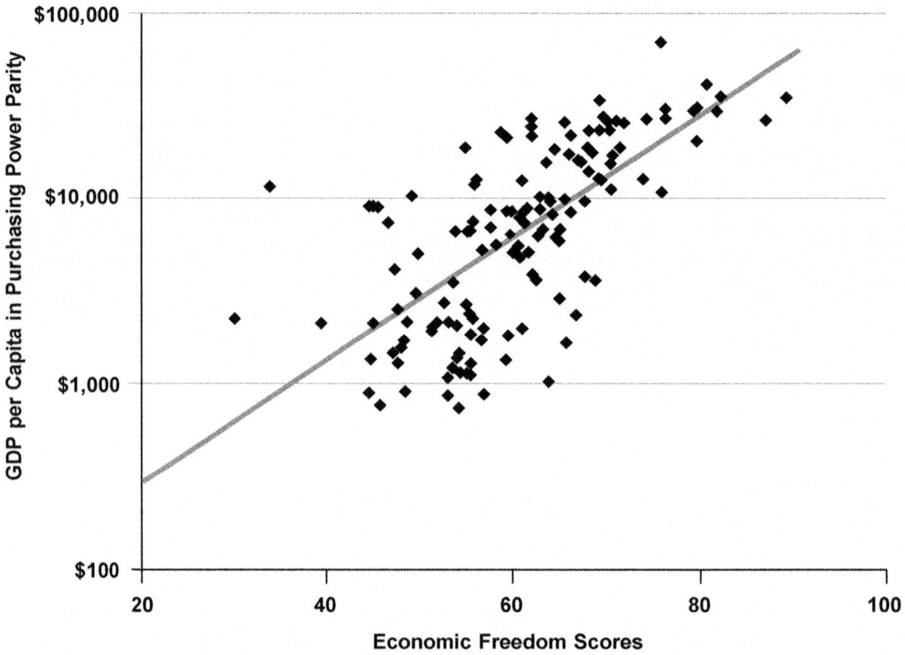

Figure 77. Free Enterprise and Per Capita Income (2009 Index of Economic Freedom)

It is important to emphasize that both the absolute levels and the changes in levels of economic freedom scores are predictors of economic growth and prosperity. Those countries that increased economic freedom grew more than a percentage point faster than those countries that lost economic freedom, averaging an annual growth rate of 3.3% in per capita GDP versus the laggards average of 1.9%.

Emphatically, over the 15-year period of assessment, those countries that gained more than 5 points of economic freedom reduced poverty over 5.8% while those countries that lost more than 5 points of economic freedom lost an additional 0.4% of their people to poverty.

It is noteworthy that the economic freedom scores also relate highly with other important human benefits:

- **Indices of democracy are correlated at +0.67, accounting for 45% of democratic governance.**
- **Indices of human development are correlated at +0.63, accounting for 40% of overall quality of life.**
- **Indices of environmental development are correlated at +0.55, accounting for 30% of environmental sustainability.**

All of this is part of a database that we have replicated and trust. A sustained commitment to economic freedom fosters human and environmental development as well as prosperity and equity of distribution.

34

Generative Entrepreneurs

President Obama has characterized small, entrepreneurial-type businesses as "ma-and-pa" operations in his mind, grossing no more than $250,000 per year. Indeed, many of the people he met on the streets of Chicago fit this description. However well this characterization may work for taxation or regulation, or even voter registration, it flat-out misses the critical contributions of entrepreneurial enterprise to the American economy. Moreover, how on earth does he reconcile this immature image of small, entrepreneurial businesses with their prepotent contributions to the economy? Under conditions of economic freedom:

- **Entrepreneurial enterprise generates as much as 80% of the scientific breakthroughs and technological innovations.**
- **Entrepreneurial enterprise generates as much as 80% of the new jobs in implementing the follow-throughs on the technological innovations.**
- **Entrepreneurial enterprise generates as much as 50% of America's GDP.**

To be sure, an overwhelming case can be made for the prepotency of entrepreneurial enterprise over all other sources of economic growth:

- **Entrepreneurial enterprise dwarfs the contributions of all governmental sources, including the presidency, congress, and the federal sector that they manage.**

Yet these entrepreneurial businesses are exploited by the multinational corporations of the private sector who, like Bill Gates of Microsoft, track them and trap them if they have attractive commercial products or services, "making them offers they can't refuse."

Moreover, they are passed over by the bureaucratic czars of the public sector who favor "sole-source contracts" with "soul-less contacts," while simultaneously taxing and regulating American entrepreneurs out of existence in the fabled American "meritocracy," "giving orders they can't follow."

To be sure, not all entrepreneurs are generative, capable of creating scientific "breakthroughs" and innovative technologies. By our estimates, no more than one in 100 entrepreneurs are currently generative; therefore somewhere around 50,000 people out of five million small businesses qualify as "generative entrepreneurs." In business, they may be labeled as "process engineers," capable of designing themselves and others out of their jobs. In government, they may be characterized as "social engineers," capable of defining spaces and empowering people to fill them.

The point is this: the other 99 of us live, learn, and work in the "draft" of their sciences, empowered by the tools of their technologies, prospering through their intelligence, their perspectives, and their good will.

In transition, America has had a unique form of capitalism, unlike any other nation. From the very founding of this country, entrepreneurs have been in the forefront. Indeed, we may say that entrepreneurs founded this country by overthrowing King George II. We may further say that these founding fathers and mothers—like John and Abigail Adams—defined this "Great Experiment" termed America.

To be sure, Adams, Franklin, and Jefferson continue to be the models for the term "entrē," or freedom of access. Even as he led the American armies, Washington fed the American armies. In the ultimate of risk-taking ventures, the outcome of the war dictated the terms of his life: if he won, he thrived; if he lost, he died.

The American economic engine has always been driven by her entrepreneurs. Great and small, they have shaped our socioeconomic progress. Known and unknown, they have generated the initiatives that brought us to where we are.

The great entrepreneurs of the 20th century come to mind: Edison (electricity), Ford (autos), Rockefeller (oil), and Morgan (finance) met regularly and shaped American policy.

Less known but more brilliant is the work of George Washington Carver. Driven by a mission to empower and free enslaved black peoples, he generated the sciences (biology), innovated the technologies (agrarian), developed the curricula (education), and distributed "best practices" (dissemination) to empower the most productive farmers in the Agrarian Age.

Webster defines entrepreneur as follows:

> **"The person who organizes, manages, and assumes the risks of a business; a successful businessman."**

While our founding fathers were certainly defined by these characterizations, the conditions of our times have generated and, indeed, demand, a new and more powerful set of requirements:

> **The person who generates an innovative idea, positions related products and services for comparative advantage in the marketplace, and capitalizes by organizing resources for productivity and profitability.**

As I sit here in the office of my expanding businesses, I wonder about the future of entrepreneurial enterprise. The data indicate that entrepreneurial enterprise is the driving engine of our economy. The data also indicate that entrepreneurial enterprise is the social ethic of our civilization. Finally, my experience suggests that there is a tangible set of learnable skills to prepare people for generative entrepreneurism.

In short, generative entrepreneurs are the most valuable players in a free American economic system. They "gin up" the economy. They empower with their progeny. They generate their legacy.

Still, they are overwhelmed by the conditions of stasis; sold out by the promises of security; overpowered by the forces of socialism.

My Takeaway

My "takeaway" from a lifetime of work in generativity solutions is that the American economy faces schizophrenic obstacles to recovery:

- As consumers, they lack employment and confidence in the market.
- As producers, they lack generativity and confidence in their ability to lead the market.

By far, the greatest obstacle is for Americans to ignite and elevate their generative functions. This means that putting the consumers back on their feet is, at best, a transitional mission. Empowering the producers with generativity vision and systems is the longer-term, overriding mission.

The "Generativity Stimulus Package" will reignite American ideation and entrepreneurial leadership. It defines the effective ingredients of "The New Economic Paradigm" that emphasizes all forms of generativity in both "nested" and synergistic interaction (see Figure 78).

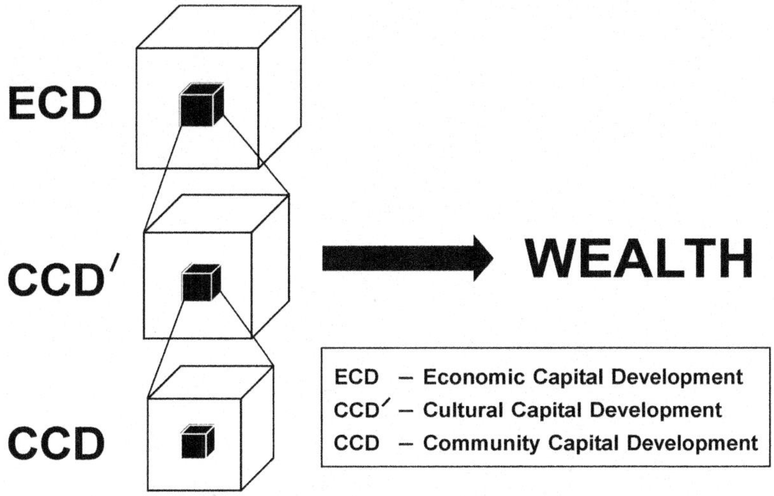

Figure 78. The New Economic Paradigm

In the final analysis, the recession of 2008–200? may be the best of all conditions for America, its citizens, and its workers. The advanced capitalist nations will weather the financial crisis better than the less mature countries. This is because capitalism, itself, is "a theory of change": whatever ingredients are most important to our prosperity are incorporated into the wealth-generating equation.

Having lost their way, Americans now have the opportunity to re-examine their values and define the requirements to accomplish these values. Are we consumers or producers? Can we return again to the generativity that made us great? Are we still worthy of the leadership role?

The generative direction is the New Capital ingredient in the economic equation. We can create our way out of any hole, no matter how deep. Moreover, we can generate a great new economy of abundance, where it is a privilege to live, learn, and, above all, work.

The overriding economic question is this: are we going to assume the economic conditions of stasis that have overtaken us over the past few decades? If so, then teach your children Mandarin and prepare them for a second-class commoditized consumer existence!

Or are we going to dedicate ourselves to the economic conditions of change—continuous change—where we generate the very conditions within which we operate? If so, then teach your children "American" and prepare them for a first-class generative producer existence!

In summary, generativity is **"the new wealth of nations."**

Generativity is the path forward for Americans.

Hopefully, others will follow.

Part X

The Next Great Civilization

35

The Conditions of Civilization

It is common in the Western world to identify the cradle of civilization as the Fertile Crescent, including especially Levant and Mesopotamia. Defining civilization as agrarian, urban, and culturally related, current thinkers have conjectured that there is no one cradle. Instead, they believe that several civilizations evolved independently, of which the Near Eastern Neolithic was the first to be identified. The enduring characterization of civilization has been the civilized way that cultures have related. This cultural relating factor applies to both homogeneous cultures such as Japan and heterogeneous cultures such as China and Asia. Other cultures have evolved through the ages reaching to open the Gordian Knot of civilization. While many have achieved hierarchical relating within cultures, few have demonstrated equitable relating between cultures and none have attained judicious relating among cultures.

CULTURAL RELATING

The Core Conditions of Civilization

The Democracy Experiments—Greece

In the history of humankind, there have been three great movements toward freedom for civilization. Each constitutes a vital experiment in the representation of human experience and the realization of human potential. Together, they defined the foundation for the Western world. The first freedom movement was embodied in the democratic government of ancient Greece in Athens. Pericles demonstrated for the world that direct democratic participation was possible. Even though flawed by the limited size of the Athenian citizenry and the exclusion of the slaves on which it was built, this demonstration was nonetheless the initial step toward a new view of the relationship between people and government. While other nations attempted to follow this ideal, expressions of democracy were neutralized by the pathology of classism derived from narcissism and its instruments of servitude (Rome and England).

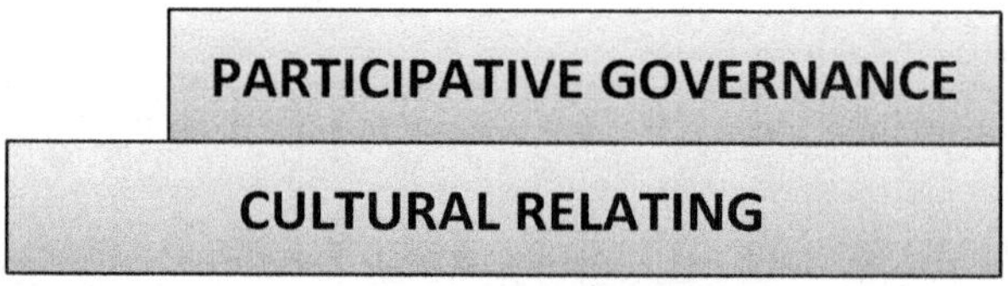

The Developmental Conditions of Democratic Civilization

The Free Enterprise Experiment—U.S.A.

The second great movement toward freedom was the American system, founded more than 200 years ago. Basing their ideas on the works of European philosophers such as Locke, Hobbes, and Hume, the Founding Fathers demonstrated that not only representative democratic governance was possible, but also that free enterprise economics was its driving engine. Of course, the American system was also flawed by withholding full freedoms and rights from first Americans, women, blacks, and other people of color. However, this flaw was not fatal. Our founders were politically conscious enough to give us an amending instrument to enact changes in the movement toward full freedom: the U.S. Constitution. The equality of all "men" having been declared, it remained for the disenfranchised to seek and achieve their own civil and human rights through amendments to the Constitution.

Indeed, today the U.S. stands alone as the beacon of democracy and free enterprise in the world:

- Its 50 states, some of them larger than most countries, relate peacefully and productively in the freest social system the world has ever known.
- Its 50 states, many of which would be among the most prosperous of countries, conduct the freest and most powerful wealth-generating economic systems in the history of the world.
- The citizens of its 50 states are constitutionally the policymakers of the freest representative democratic governance of the most powerful nation in the history of the world.

The U.S. has truly demonstrated itself to be the "Great Experiment." And it has been validated by all known measurements!

The Culminating Conditions of Free Enterprise Civilization

The New Capitalism Experiment—Global

Defining new capital as "what is most important," the third freedom movement has generated the greatest waves of prosperity in the history of humankind. Over the past 50 years, the new capitalism has out-generated the cumulative wealth of our entire history. To be sure, in the process, the new capitalism has made every individual, every community, every culture, every nation, more prosperous than they were before 1960.

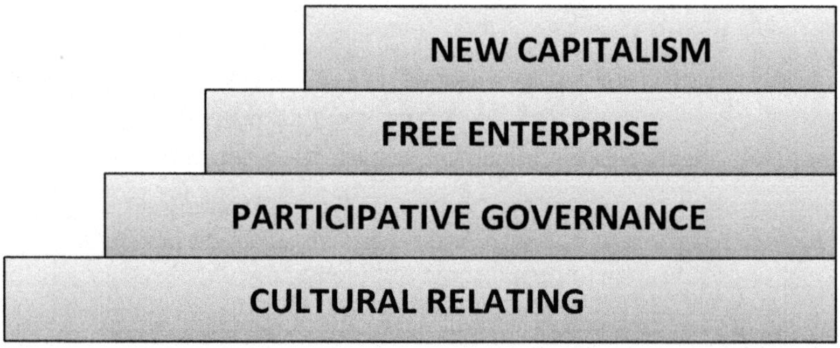

The Cumulating Conditions of the New Capitalism

The Generativity Experiment

Evolving in our times, the potential for the fourth freedom movement is projected by the Generativity Doctrine, which offers a futuristic vision of the global village and its marketplace. Informed by the spiraling achievements of the American experiment, the Generativity Doctrine points civilization toward an elevated and integrated global society based on the fundamental proposition of Global Freedom: "All people are created equal in their potential for empowerment in generativity." The functions of such generativity incorporate the culminating conditions of generativity:

- Free and interdependent cultural relating
- Free and fully participative governance
- Entrepreneurially driven, free enterprise economics

The methods for achieving these freedom functions tell the story of the Generativity Doctrine.

We are thus called to engage in the Next Great Experiment—to build our people, our communities, all cultures, and all nations into an elevated, interdependent and, above all else, generative global society. This is the Generativity Doctrine. It will generate the Next Great Civilization! And it will be validated again and again!

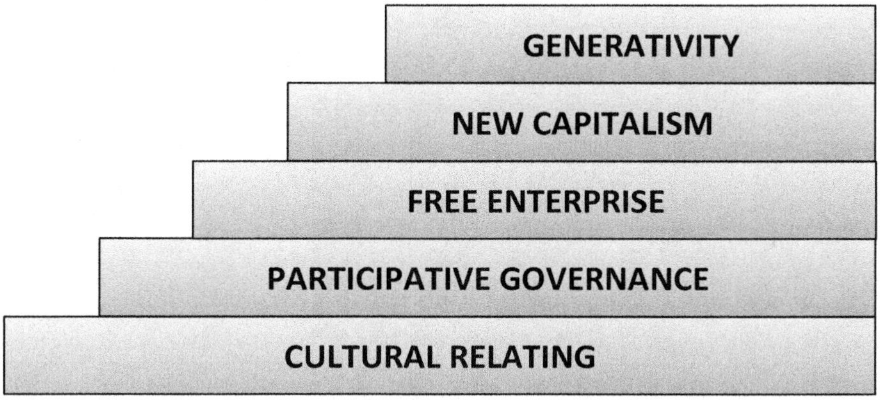

The Explosive Conditions of Generativity

36

The Next Great Civilization—??

The next great vision is the image of the socioeconomic growth mission responsive to the needs of all its citizens. It is humankind's greatest hope to generate a vision of the ideal and then proceed to live inside this vision. In our brief time here, the American people have done this many times:

- Creating an exemplary agrarian society in a natural world
- Giving leadership to an industrial civilization in an agrarian world
- Generating a new and bold information civilization in an industrial world

Each stage of evolution has "nested" the previous stage. Indeed, each stage has actualized the previous stage. This is the contribution of American civilization: continuous change to fulfill the socioeconomic growth ethic of the Next Great Civilization. Who will lead? Who will follow?

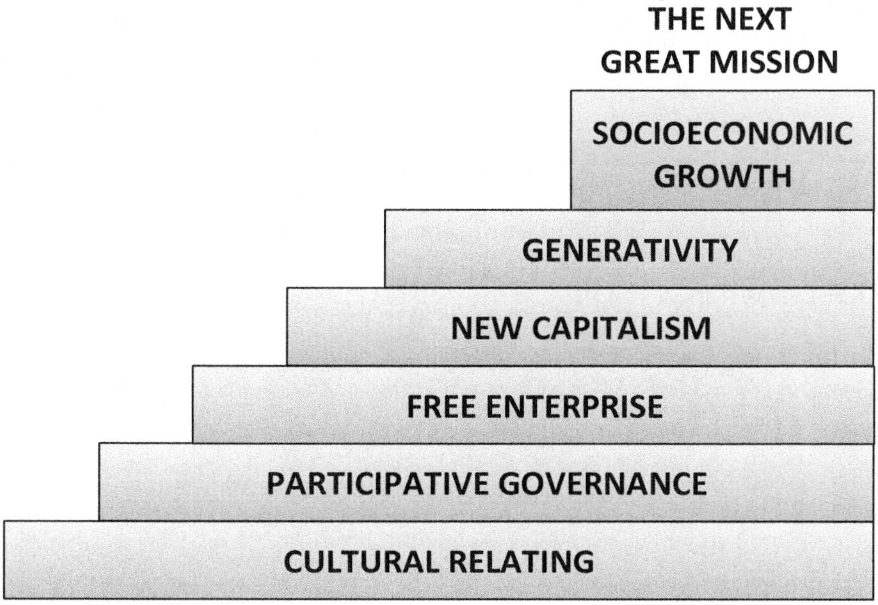

Accomplishing the Mission of the Next Great Civilization

Socioeconomic Growth Mission

We may view the socioeconomic growth mission in sharpest relief by again referring to the Socioeconomic Growth Paradigm (see Figure 79). This is the image that defines the outcomes of the past 50 years driven by the New Capitalism:

- It demonstrates the prodigious prosperity generated by our growth conditions.
- It also demonstrates the inequity of distribution that has burdened our accomplishments.

The resolution of the "Growth versus Distribution Conflict" will occur only with the maturity of the "Next Great Civilization." Clearly, only the universality of empowerment in generativity will create the facilitative conditions for resolving this potential crisis. The outcomes will be generated by respect for "The Reciprocity Value."

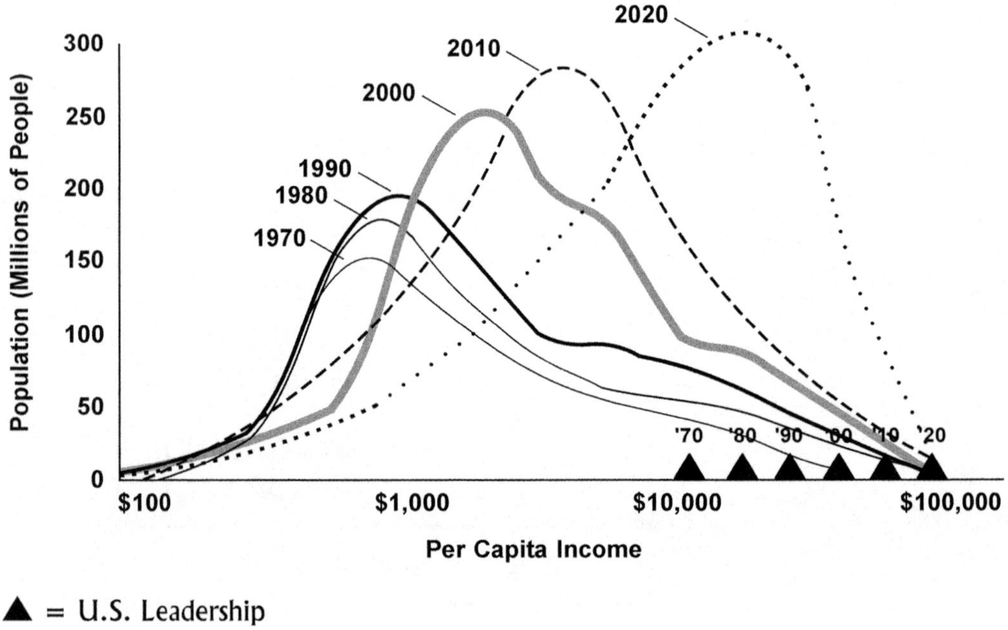

Figure 79. The Socioeconomic Growth Paradigm

The Reciprocity Value

Paramount among the values of the Next Great Civilization is the requirement of reciprocity or, in more spiritual terms, "The Golden Rule: Do unto others as you would have them do unto you." The civilization requires personal responsibility for the freedom and rights it offers. The ultimate reciprocity is mutual: each one's values become the other's requirements. Succinctly, just as civilization's values become the citizen's requirements, our values become the civilization's requirements. These are the functions of free choice in the Freedom Matrix (see Figure 80).

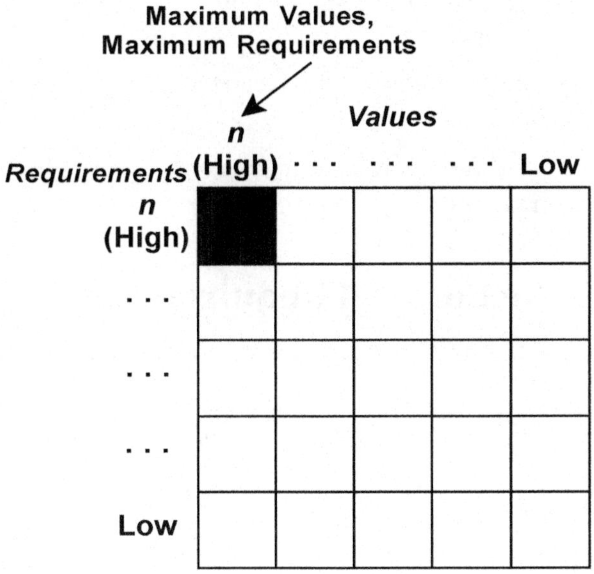

Figure 80. The Freedom Matrix of a Great Civilization

The Civilization's Values

Of course, the Next Great Civilization has evolving values of its own. Generically, it follows the principles of possibilities science:

- To relate to her people and accurately obtain their frames of reference on all relevant issues
- To empower her people so that they can relate and initiate as enlightened citizens to accomplish their own purposes
- To free or release her empowered people to seek and achieve their own visions of their civilization and their universes

While the Next Great Civilization's values become her citizen's requirements, each individual citizen will be freed to create his or her own world.

The Civilized Requirements

For example, the Next Great Civilization's values are transformed into evolving community requirements:

- Every child shall have the opportunity to be reared by generative parents or surrogates.
- Every student shall be taught by generative teachers.
- Every learner shall be empowered in generativity thinking skills.
- Every public sector government agency and its representatives shall be driven by generativity services.
- Every private sector company and its leaders shall generate incessantly new and elevated economies.

Again, these may be considered evolving requirements that are to be processed and transformed into citizen values.

The Cultural Requirements

Similarly, the Next Great Civilization generates cultural requirements:

- Every culture and every individual within every culture shall relate to members of every other culture by relating to their experience and generating on their behalf.
- Every citizen who is the member of any group shall relate and generate respectfully with any other groups on all issues affecting or being affected by them.
- Every worker at any level of any company or agency shall generate entrepreneurially or intrapreneurially to accomplish socioeconomic prosperity on behalf of all.

Once again, these may be considered evolving requirements that are to be processed or transformed into cultural and/or personal values.

The Socioeconomic Requirements

Finally, the Next Great Civilization generates socioeconomic requirements:

- Every productive component of the socioeconomic marketplace shall relate to every other component:
    - The generators shall relate to the innovators and both shall relate to the commercializers who constitute the "GIC Triad" that makes the market "go 'round."

- The "GIC Triad" shall relate to the components of the "Cultural Triad" in order to make the world "go 'round": cultural relating, participative governance, free entrepreneurial enterprise.
- The "GIC Triad" and the "Cultural Triad" shall relate to the community in order to serve and be served by the people who make everything "go 'round": home, school, college, governance, business.

Again, the theme of all components of socioeconomic growth is interdependent generativity.

Socioeconomic Growth

We now generate an operational scale for s**ocioeconomic growth.** The levels are developmental and cumulative. In this book, we have made the case for the contribution of each of these levels to the **Socioeconomic Growth Mission** (see Table 10). In simplistic terms, the mission is to make the pie large enough for all to share and to empower everyone to generate a piece that responds to their needs and hopes.

Table 10.
Socioeconomic Growth

LEVELS	FUNCTIONS
5	**Generativity**
4	**New Capitalism**
3	**Free Enterprise**
2	**Participative Governance**
1	**Cultural Relating**

The Generation of Resources

We may **"factor"** our socioeconomic functions in terms of resource generation and distribution (see Table 11). As may be viewed, each level of development provides the platform for the next level. Moreover, each level of development fulfills the potential of the previous level. In this context, all socioeconomic functions are currently nested in the **"tsunami wave"** produced by **"The New Capitalism."** In the future, all functions will be nested in **"The Generativity Doctrine."**

Table 11.
Socioeconomic Growth

LEVELS	FUNCTIONS		RESOURCE DEFINITIONS
5	**Generativity**	—	Generating New Forms of Resources
4	**New Capitalism**	—	Producing New Resources
3	**Free Enterprise**	—	Trading Resources
2	**Participative Governance**	—	Allocating Resources
1	**Cultural Relating**	—	Sharing Resources

The Generativity Paradigm

Once again, "generativity is the human brainpower necessary to create new and more productive ideas."

It means "seeing beyond the high beams" and, indeed, "beyond the moon beams."

After researching the dimensions of generativity, we are now able to empower individuals, organizations, communities, cultures, and economies to apply and transfer the generativity processes (see Figure 81).

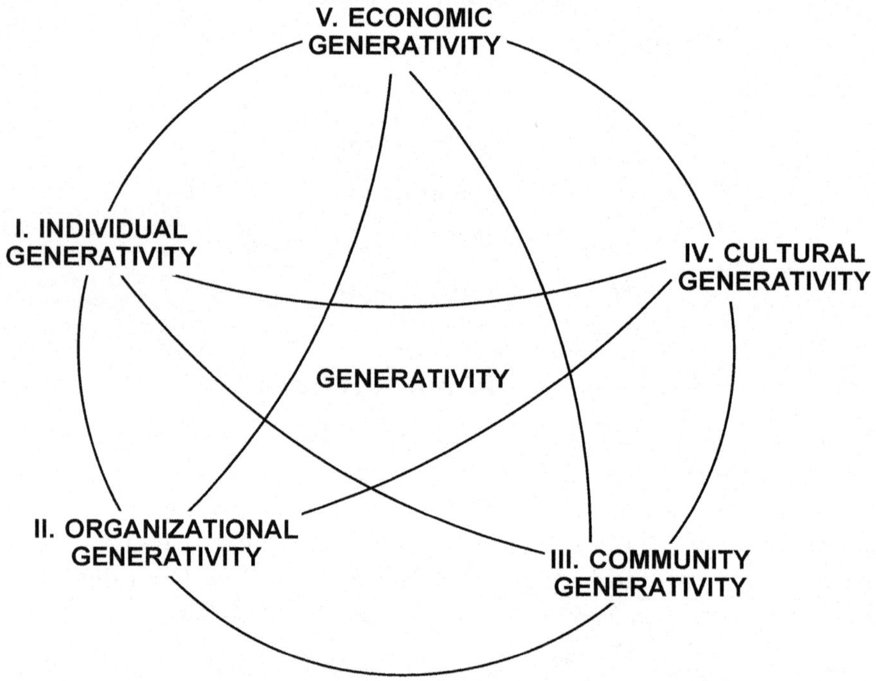

Figure 81. The Generativity Paradigm

Your Values

Impossible as it may sound, you can become whatever you wish to be. Directed by the vision of generativity, empowered by its technologies in all areas—individual, informational, organizational, community, cultural, socioeconomic—you can take the simplest steps toward becoming enlightened citizens in a generativity civilization:

- Think of your core values by which you have guided your life.
- Think of your most ideal instance of your core value.
- Define the ideal "operationally" in terms of the operations needed to achieve the ideal.

Now you are prepared to embrace your own responsibilities for achieving your ideal.

Your Requirements

Every great parent reminds us of the responsibilities that accompany our choices. So our choices are truly not free! Our responsibilities are the operations that we must perform in order to accomplish our goals (see Figure 82).

1. Throw out a "sky hook" to define our ideal goals.
2. "Pick ourselves up by our bootstraps" to develop our programs.
3. "Take our first real steps."

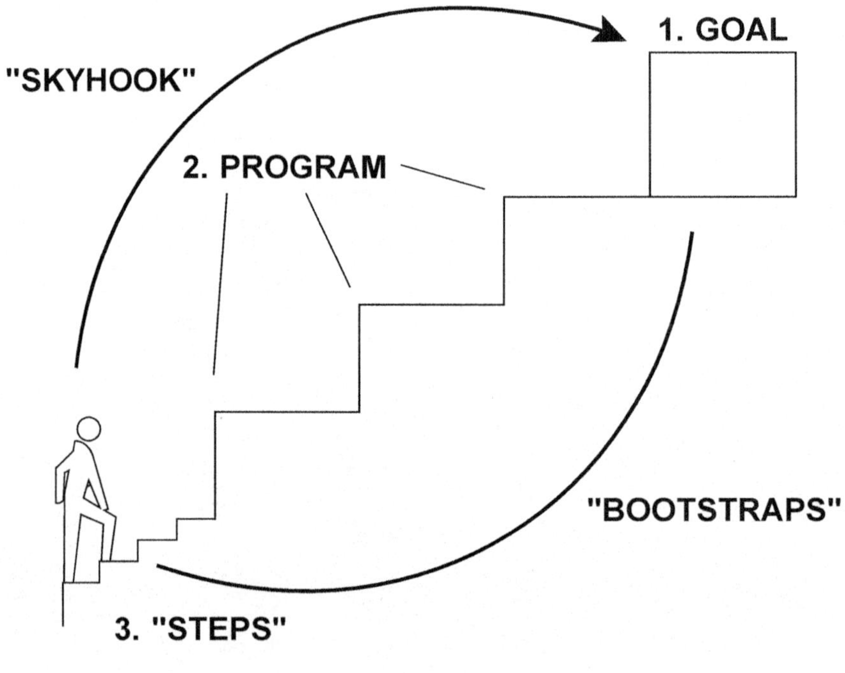

Figure 82. The Personal Generativity Architecture

This is the "Personal Generativity Architecture" to accomplish any ideal vision.

Who Are We?

The Lenni-Lenape Indians are famous for having the oldest written word in North America. It is a tribal record carved on reeds. Labeled the "Walam Olum," the record spoke of their first encounter with the alien intruders:

> ***"People floating in from the east. The whites are coming...who are they?"***

In 1682, Tammany, chief of the Lenape, characterized William Penn and others with whom he negotiated a treaty:

> ***"If so few can come so far and do so much, how much can many do?"***

Many years later, regretful of the treaties that had not been reciprocated and the lands that had been confiscated, Tammany repaired to his wigwam in what is now Princeton and immolated himself and his homestead. Perhaps he had discovered who we were!

Appendices

- **Afterword: Psychenomics**
  — Richard C. Sprinthall
- **References**
- **Carkhuff Body of Work**

Afterword: Psychenomics

Richard C. Sprinthall, Ph.D.

mcleanproject.com

Psychenomics

When we invited Robert R. Carkhuff and Andrew H. Griffin to "Save Springfield," it was, to be sure, a last-resort measure. It is true that Carkhuff did bring with him an extensive research-based knowledge of individual and organizational generativity. In other words, he knew how to make people and organizations work effectively and, moreover, innovatively.

However, we had no idea of the awesome community benefits we would receive. Least of all, I had no idea of the scientific discipline Carkhuff and associates would establish: "Psychenomics"—the application of the science of psychology to the practice of economics. Here is how I now understand this new discipline.

First of all, Carkhuff was troubled by the limitations of traditional "Probabilities Science" and its measurement tools, "Parametric Statistics." The traditional functions of science—describing, predicting, and controlling—dictated the architecture for construction and policing of a ghetto from the top-down: totalitarian governance, control economics, and dependency relating.

Confronted with these unacceptable consequences, Carkhuff and his life-long processing partner, Dr. Bernard G. Berenson, generated "The New Science of Possibilities" with its possibilistic functions—Relating, Empowering, Freeing—which, in turn, generated the building of the community from the bottom-up: individual, organizational, community, cultural, and economic development.

Carkhuff and Griffin made a pact: Griffin would represent the minority community and Carkhuff would represent the majority community, including especially the corporate community with whom the minority community so desperately needed a relationship. Their pact was interdependent and synergistic: the community would grow as the corporate community grew; the corporate community would grow as the minority community grew.

The different sub-communities involved were held together by "the cellular bonding tissue" of cultural generativity: interdependent cultural relating, participative democratic governance, and entrepreneurial capitalistic enterprise.

In turn, these different sub-communities were dedicated increasingly toward economic generativity: generating, innovating, and commercializing in interdependent and synergistic relationships.

Lo and behold, in the 1970s and 1980s, Springfield was transformed dramatically from its declining manufacturing base to an exemplary information technology center served by numerous technological boutiques.

However, as a social scientist, I now stand in awe of the new science that Carkhuff and his associates have discovered. Here are the salient features of Psychenomics:

- It employs the full power of Probabilities Science and related Parametric Measurement to describe and predict all phenomena.

- It generates the full power of Possibilities Science and its non-Parametric Measurements to relate, empower and free the full potential of all phenomena and, in so doing, directs the operations of Probabilities Science.
- It employs the prepotent power of Generative Human Processing—individual, organizational, community, cultural, economic and a new "Paradigmetric Measurement System" to represent the universal processing of all information—conceptual, operational, dimensional, vectorial, phenomenal.

Psychenomics is a true science—generative, measureable, and replicable. It explicates the unknown by describing, predicting, relating, empowering and releasing or freeing phenomena. Economics is simply another instance of the application to human endeavor.

In terms of systems thinking, here is a simple example of how Carkhuff and Psychenomics diagnosed our current economic crises (see Figure 83):

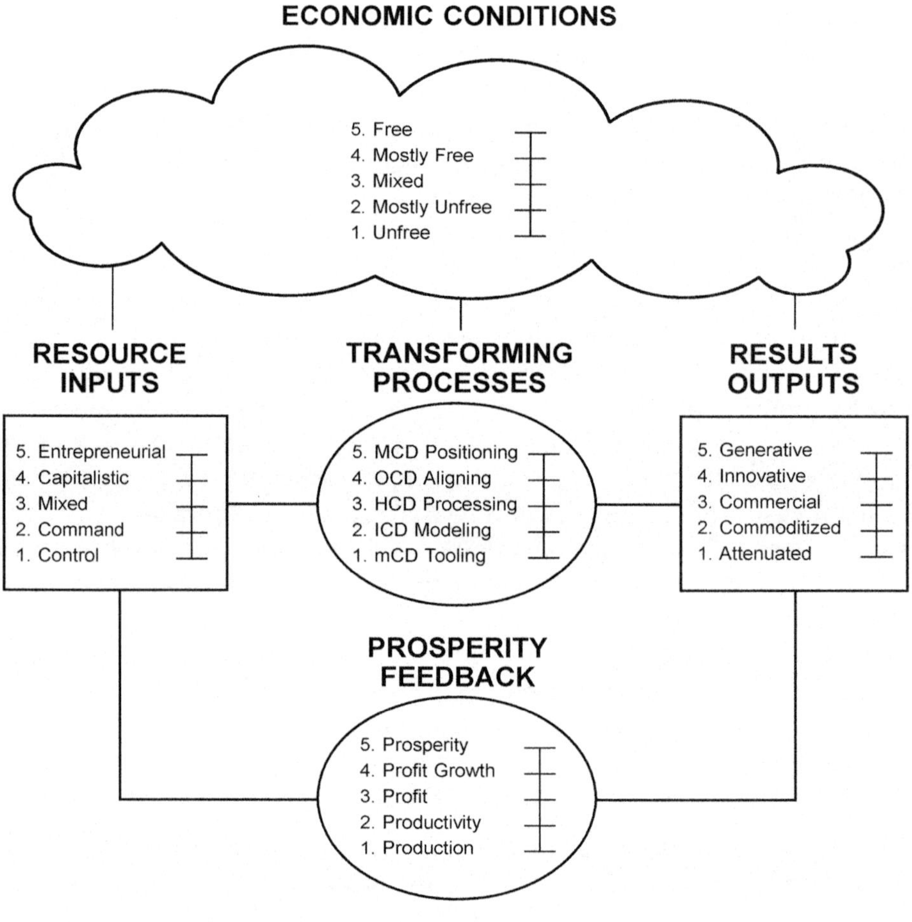

Figure 83. Systems Analyses of Carkhuff's "Psychenomics"

- **Economic Conditions**—The economic conditions determine up to 50% of our prosperity. At the highest levels of freedom, people can generate, innovate and commercialize. At the lowest, unfree levels, people are unable to process in any way. Carkhuff, himself, was offered the highest levels of economic freedom in his architecture and implementation for "Saving Springfield."
- **Resource Inputs**—The entrepreneurial enterprisers are the primary resource inputs. Potentially, they may contribute the other 50% of the variance in our prosperity. At the highest levels of enterprise, entrepreneurs engage the economic conditions in an attempt to generate, innovate, and commercialize. At the lowest levels of enterprise, chain-of-command organizations dictate programmed performances in command-and-control economics. Carkhuff, himself, was offered carte blanche support in architecting and engineering "Saving Springfield."
- **Transforming Processes**—The actualization of the entrepreneur's contribution to prosperity is contingent upon the interaction with the transforming processes. The first issue is the inclusiveness of the entrepreneur's empowerment with the levels of "New Capital Development (NCD)":
  - **MCD**—Marketplace Positioning
  - **OCD**—Organizational Alignment
  - **HCD**—Human Processing
  - **ICD**—Information Modeling
  - **mCD**—Mechanical Tooling

  In Carkhuff's approach to "Saving Springfield," he operationally defined and systematically implemented all of the NCD technologies.
- **Results Outputs**—Likewise, the actualization of the entrepreneurial contribution to prosperity is contingent upon the relationships to results outputs: generative, innovative, commercial, commoditized, attenuated, or some combination of these outputs. All levels of the transforming processes (MCD, OCD, HCD, ICD, mCD) are potentially involved. Positioning is critical: Do we meet the requirements of our positioning with our capabilities? In Carkhuff's case, he positioned Springfield "Generative–Innovative": Generative with the Possibilities architectural "breakthroughs," and innovative with the Probabilities systems follow-through.
- **Performance Feedback**—Finally, the analysis culminates with the measurement of the prosperity feedback:

5. Prosperity $\left(\frac{\$\,\mathbf{RO} >}{\$\,\mathbf{RI}}\right)^{n} > \left(\frac{\$\,\mathbf{RO} >}{\$\,\mathbf{RI}}\right)^{\cdots}$

4. Profit Growth $\left(\frac{\$\,\mathbf{RO} >}{\$\,\mathbf{RI}}\right)^{2} > \left(\frac{\$\,\mathbf{RO} >}{\$\,\mathbf{RI}}\right)^{1}$

3. Profitability $\frac{\$\,\mathbf{RO} >}{\$\,\mathbf{RI}}$

2. Productivity $\frac{\mathbf{RO} >}{\mathbf{RI}}$

1. Production $\frac{\mathbf{RO}}{\mathbf{RI}}$

The success of all levels of feedback are contingent upon the relationship between results outputs (RO) and resource inputs (RI): economic growth is predicated upon getting more value in RO than we invested in RI. In "Saving Springfield," Carkhuff generated high levels of prosperity that were universally identified as "The Springfield Miracle."

In summary, the scaled measurements of Psychenomics contrast vividly with the current contradictory and conceptual ideologies. Throughout all of his generativity demonstrations, Carkhuff has maintained his goal of demystifying the operations behind the words. Thus, he delivers data-based policies for "Saving America" just as he did for "Saving Springfield": Elevate our freedoms! Empower our generativity! Turn Entrepreneurial Enterprisers loose!

In conclusion, the power of Psychenomics was demonstrated by the generativity of its creator. Carkhuff is the living model of "The Generative Entrepreneur." Psychenomics is his legacy to a **Generativity Civilization—Prosperous, Participative, Peaceful!**

Richard C. Sprinthall, Ph.D.
Professor Emeritus,
American International College

January 2013
Springfield, MA

References

Anthony, W. *The Principles of Psychiatric Rehabilitation.* Baltimore, MD: University Park Press, 1979.

Aspy, D. N. and Roebuck, F. N. *Kids Don't Learn from People They Don't Like.* Amherst, MA: HRD Press, 1977.

Banks, G. *From Bondage Through Prosperity: Finding the Freedom in Thinking.* Amherst, MA: HRD Press, 2013.

Berenson, B. G. *The Possibilities Mind.* Amherst, MA: HRD Press, 2001.

Berenson, B. G. *Carkhuff and the Human Sciences.* McLean, VA: The McLean Project, 2013.

Berenson, B. G. and Cannon, J. R. *The Science of Freedom.* Amherst, MA: HRD Press, 2006.

Berners-Lee, T. *Weaving the Web.* Britain: Orion Business, 1989, ISBN 0-7528-2090-7.

Bierman, R. *Toward Meeting Fundamental Human Service Needs.* Guelph, Ontario: Human Service Community, Inc., 1976.

Bugelski, B. R. *Psychology of Learning.* New York: Holt, Rinehart & Winston, 1956.

Bugelski, B. R. *Principles of Learning.* New York: Praeger, 1979.

Carkhuff, C. J. and Paley, G. *IONN—Interdependent Object Networking and Navigation System.* McLean, VA: GenStar, 2012.

Carkhuff, C. J. and Paley, G. *VOR—Virtual Organization Reality.* McLean, VA: GenStar, 2012.

Carkhuff, R. R. *Helping and Human Relations. Volumes I and II.* New York: Holt, Rinehart & Winston, 1969.

Carkhuff, R. R. *The Development of Human Resources.* New York: Holt, Rinehart & Winston, 1971.

Carkhuff, R. R. *The Promise of America.* Amherst, MA: HRD Press, 1974.

Carkhuff, R. R. *Toward Actualizing Human Potential.* Amherst, MA: HRD Press, 1981.

Carkhuff, R. R. *Sources of Human Productivity.* Amherst, MA: HRD Press, 1983.

Carkhuff, R. R. *The Exemplar.* Amherst, MA: HRD Press, 1984.

Carkhuff, R. R. *Human Processing and Human Productivity.* Amherst, MA: HRD Press, 1986.

Carkhuff, R. R. *The Age of the New Capitalism.* Amherst, MA: HRD Press, 1988.

Carkhuff, R. R. *Empowering.* Amherst, MA: HRD Press, 1989.

Carkhuff, R. R. *Human Possibilities.* Amherst, MA: HRD Press, 2000.

Carkhuff, R. R. *The Age of Ideation.* Amherst, MA: HRD Press, 2007.

Carkhuff, R. R. *The Art of Helping.* Ninth Edition. Amherst, MA: HRD Press, 2009.

Carkhuff, R. R. *Saving America: The Generativity Solution.* Amherst, MA: HRD Press, 2010.

Carkhuff, R. R. *The Human Sciences: Volume I. Probabilities, Possibilities, and Generativity Sciences.* Amherst, MA: HRD Press, 2013.

Carkhuff, R. R. *The Human Sciences: Volume II. Probabilities, Possibilities, and Generativity Technologies.* Amherst, MA: HRD Press, 2013.

Carkhuff, R. R. *Human Generativity.* Amherst, MA: HRD Press, 2013.

Carkhuff, R. R. www.mcleanproject.com. The McLean Project, 2011.

Carkhuff, R. R. and Berenson, B. G. *The New Science of Possibilities. Volumes I and II.* Amherst, MA: HRD Press, 2000.

Carkhuff, R. R. and Berenson, B. G., et al. *The Possibilities Organization.* Amherst, MA: HRD Press, 2000.

Carkhuff, R. R. and Berenson, B. G., et al. *The Possibilities Leader.* Amherst, MA: HRD Press, 2000.

Carkhuff, R. R. and Berenson, B. G., et al. *The Freedom Doctrine.* Amherst, MA: HRD Press, 2003.

Carkhuff, R. R. and Berenson, B. G., et al. *Freedom-Building.* Amherst, MA: HRD Press, 2003.

Carkhuff, R. R. and Berenson, B. G., et al. *The Freedom Wars.* Amherst, MA: HRD Press, 2005.

Carkhuff, R. R. and Berenson, B. G., et al. *The Possibilities Economy.* Amherst, MA: HRD Press, 2006.

Drasgow, J. Eclipsing All Great Works. Foreword, *The Freedom Wars.* Amherst, MA: HRD Press, 2000.

Einstein, A. *Relativity: The Special and General Theory.* New York: Henry Holt, 1931.

References

Einstein, A. *The Evolution of Physics.* Cambridge: University of Cambridge, 1938.

Einstein, A. *Collected Papers of Albert Einstein.* Princeton, NJ: Princeton University Press, 1989.

Hull, C. L. *Mathematics—Deductive Theory of Rote Learning.* New York: Appleton–Century–Crofts, 1940.

Hull, C. L. *Principles of Behavior.* New York: Appleton–Century–Crofts, 1943.

Hull, C. L. *A Behavior System.* New Haven, CT: Yale University Press, 1952.

Kakovitch, T. *The Fifth Force.* Amherst, MA: HRD Press, 2012.

Kakovitch, T. *Collegium.* McLean, VA: The McLean Project, 2012.

Kilby, J. *First Successful Demonstration of Integrating a Transition with Resistors and Capacitors on a Simple Semiconductor Chip Defining the Monolithic Idea.* Dallas, TX: Texas Instruments, September 12, 1958.

Pavlov, I. P. *Conditioned Reflexes.* Oxford: Oxford University Press, 1927.

Rogers, C. R. The Necessary and Sufficient Conditions of Therapeutic Personality Change. *Journal of Consulting Psychology,* 1957, 22, 95–103.

Siegel, S. *Nonparametric Statistics for the Behavioral Sciences.* Washington, DC: American Association for the Advancement of Science, 1959.

Sprinthall, R. C. *Basic Statistical Analysis.* Boston, MA: Allyn and Bacon, 2012.

Sprinthall, R. C. *Psychenomics.* Afterword in R. R. Carkhuff, *Saving America: The Generativity Solution.* Amherst, MA: HRD Press, 2010.

Sprinthall, R. C. *SPSS.* Boston, MA: Pearson Education, Inc., 2009.

Straus, E. *Phenomenology: Pure and Applied.* Pittsburg: Duquesne University Press, 1964.

Truax, C. B. and Carkhuff, R. R. *Toward Effective Counseling and Therapy.* Chicago, IL: Aldine, 1967.

Watson, J. *Behaviorism.* Chicago, IL: University of Chicago Press, 1930.

Carkhuff Body of Work

Robert R. Carkhuff, Ph.D.

The Generativity Solution:
Science in the Service of Humankind

An Annotated Body of Work

CONTENTS

1. Introduction and Overview

Carkhuff was 15 years out of graduate studies in psychology when the world of science first took note of his scientific contributions in the study of helping and human relations.

Today, Carkhuff is the most powerful force in the history of science representing the science and technology of **"human generativity"** and formulating **"The Human Sciences."**

With a profound commitment to establishing a **"True Science of Human Behavior,"** Carkhuff has been the path-finding leader for the past 50 years as his body of work will testify.

In 1978, in terms of frequency of citations in psychology, Carkhuff ranked among the leading contributors (see Table 1). Two volumes of ***Helping and Human Relations*** and one volume on ***Counseling and Psychotherapy*** led the listings.

Table 1.
Most-Cited Books in Clinical Psychology [1]

Bandura, A. *Principles of Behavior Modification,* 1969.
Carkhuff, R. R. *Helping and Human Relations, Volumes I and II,* 1969.
Kelly, G. A. *The Psychology of Personal Constructs,* 1955.
Truax, C. B. and Carkhuff, R. R. *Toward Effective Counseling and Psychotherapy,* 1967.
Fenichel, O. *The Psychoanalytic Theory of Neurosis* (1945)
Freud, S. *Zur Geschichte der Psychoanalytischen Bewegung* (On the History of the Psychoanalytic Movement), 1914.

2. Helping and Human Relations

Carkhuff led the revolution of the helping professions from theoretical to operational treatment in the late 1960s. He and his associates defined the effective ingredients of helping in operational terms:

- *Toward Effective Counseling and Psychotherapy.* Chicago, IL: Aldine, 1967 (with C. B. Truax).
- *Sources of Gain in Counseling and Psychotherapy.* New York: Holt, Rinehart & Winston, 1967 (with B. G. Berenson).
- *Beyond Counseling and Therapy.* New York: Holt, Rinehart & Winston, 1967.

Basically, the helping dimensions, such as empathy and respect, facilitated helpee exploration leading to indices of therapeutic personality change. The book with Truax (Aldine, 1967) was listed among the most-cited books in clinical psychology. [2]

Carkhuff extended these core findings to all helping and human relations. Simultaneously, he developed training programs for learning these operational skills in the practice of helping:

- *Helping and Human Relations. Volume I: Selection and Training.* New York: Holt, Rinehart & Winston, 1969
- *Helping and Human Relations. Volume II: Practice and Research.* New York: Holt, Rinehart & Winston, 1969
- *The Art of Helping.* Amherst, MA: HRD Press, 1971

Operationally, the helper's skills were refined to emphasize responding, personalizing, and initiating in order to facilitate the helpee's process, which was expanded to incorporate exploring, understanding, and acting, all of which led to therapeutic change. Now in its 9th edition, *The Art of Helping* has sold more than one million copies. In turn, the two volumes of *Helping and Human Relations* were among the most-cited books in social sciences. [3]

Along with his three books, Carkhuff was, himself, identified among the most-referenced social scientists by *The Institute of Scientific Information.* [4] [5] [6] Altogether, these works culminated the operationalization of previously theoretical processes.

3. Educational and Community Applications

In the 1970s and 1980s, Carkhuff transferred his findings to human and community development in the private as well as public sectors:

- *The Development of Human Resources: Education, Society and Social Action.* New York: Holt, Rinehart & Winston, 1971
- *Toward Actualizing Human Potential.* Amherst, MA: HRD Press, 1981
- *The Exemplar: The Exemplary Performer in the Age of Productivity.* Amherst, MA: HRD Press, 1983
- *Sources of Human Productivity.* Amherst, MA: HRD Press, 1984

Operationally, Carkhuff defined the productive human performer in the productive organizational system that, in turn, is defined in the productive community development system.

John T. Kelly, Director Emeritus, Advanced Systems Design, IBM, Inc., offered this evaluation of Carkhuff's work:

> **"Carkhuff offers us a vision of the future. It is a vision of a great Age of Productivity, an age in which the human products and services are effectively increased so that all people can share. It is a vision of an age in which the resource inputs, natural and otherwise, are effectively invested so that no people are deprived of their birthrights."**
>
> (Kelly, Foreword, *Sources of Human Productivity,* 1984, p. xii)

During this period, Carkhuff and his associates launched a series of training products in teaching, training, and instructional systems design:

- *The Skills of Teaching Series, Volumes I–VI.* Amherst, MA: HRD Press, 1977–1981
- *Instructional Systems Design, Volumes I and II.* Amherst, MA: HRD Press, 1984
- *Training Delivery Skills, Volumes I and II.* Amherst, MA: HRD Press, 1984

Systematically, these works broke teaching and training down into four skill sets: interpersonal skills, content development skills, lesson planning skills, teaching delivery and classroom management skills. The educational initiatives culminated in the issue of the journal, *Education,* dedicated to Carkhuff. [7]

4. Human and Organizational Processing

In the late 1980s, Carkhuff created systematic skills for human processing or thinking with applications for the development of **Human and Information Capital:**

- *Interpersonal Skills and Human Productivity.* Amherst, MA: HRD Press, 1984
- *Human Processing and Human Productivity.* Amherst, MA: HRD Press, 1986
- *The Age of the New Capitalism.* Amherst, MA: HRD Press, 1988
- *Empowering: The Creative Leader in the Age of the New Capitalism.* Amherst, MA: HRD Press, 1989

Carkhuff's mentor, B. R. Bugelski, a protégé of Clark Hull, one of the founders of American Psychology, commented as follows:

> **"This work rationalizes all of our efforts in learning theory and promises the culmination of psychology in a science of processing."**
>
> (Bugelski, Review, *Human Processing and Human Productivity,* 1984)

Carkhuff summarizes the research of the effects of individual, interpersonal, and organizational processing systems in hundreds of studies of more than 150,000 recipients. Basically, Carkhuff defined Human Capital Development or HCD as generative thinking. Carkhuff's work was reviewed by the distinguished social scientist, C. H. Patterson, University of Illinois:

> **"This revolution has an important social significance also. I have stated elsewhere that the extent to which a society and its institutions, including its economic systems, facilitate the development of self-actualizing persons constitutes the criterion for the goodness of that society. To the extent that our society incorporates Carkhuff's system, it will become a better society for all its members."**
>
> (Patterson, Foreword, *Interpersonal Skills and Human Productivity,* 1983, p. 5)

5. Probabilities and Possibilities Sciences

In the 1990s and 2000s, Carkhuff, Berenson, and associates introduced the "Science of Possibilities" to drive the historical "Science of Probabilities":

- *The New Science of Possibilities, Volume I. The Processing Science.* Amherst, MA: HRD Press, 2000
- *The New Science of Possibilities, Volume II. The Processing Technologies.* Amherst, MA: HRD Press, 2000
- *Human Possibilities.* Amherst, MA: HRD Press, 2000
- *The Possibilities Leader.* Amherst, MA: HRD Press, 2000
- *The Possibilities Organization.* Amherst, MA: HRD Press, 2000
- *The Possibilities Economy.* Amherst, MA: HRD Press, 2005

Dr. David N. Aspy, renowned scientist, educator, and protégé of Robert Oppenheimer, commented on Carkhuff's contribution to the advancement of civilization:

> **"To support his views, Carkhuff does not simply offer up the science of science. He also presents the most exhaustive body of research and demonstration on relating and empowering ever presented in behavioral science. Moreover, he engaged in the most advanced demonstrations of phenomenal potential, including human, ever attempted...**
>
> **"Together, these process-centric breakthroughs will lead us to a grand new Age of Enlightenment—The Age of Ideation, and in the process, The Science of Freedom."**
>
> (Aspy, *Window on the Universe, The Science, of Freedom* 2007, p. 224)

6. New Capitalism and Freedom-Building

Eighteen months after having been introduced to Carkhuff's book, *"The Age of the New Capitalism,"* (1988), Pope John Paul II issued his Papal Encyclical, **"The New Capitalism."** In it, the Pope conceded the prepotent power of **"human generativity"** and committed himself to the **"free economy"** as the only alternative after the failure of Communism:

> Can it perhaps be said that, after the failure of Communism, capitalism is the victorious social system, and that capitalism should be the goal of countries now making efforts to rebuild their economy and society?
>
> If by "capitalism" is meant an economic system that recognizes the fundamental and positive role of business, the market, private property, and the resulting responsibility for the measure of production, as well as free human creativity in the economic sector, then the answer is certainly in the affirmative, even though it would perhaps be more appropriate to speak of a "business economy," "market economy," or simply "free economy." (John Paul II, **Centesimus Annus,** 42.1–42.2, in Miller, ***Encyclicals***)
>
> **Source:** Carkhuff, R. R. *The Freedom Wars.* HRD Press, 2004

Early in the 21st century, Carkhuff and associates introduced the models and systems for freedom-building:

- *The Freedom Doctrine.* Amherst, MA: HRD Press, 2003
- *Freedom-Building.* Amherst, MA: HRD Press, 2004
- *The Freedom Wars.* Amherst, MA: HRD Press, 2004

In evaluating the contributions of Carkhuff to the advancement of civilization, John R. Cannon, Chief Executive Officer, Human Technology, Inc., summarizes Carkhuff's contributions [8] [9]:

> **"Together, these works represent the contributions of The Possibilities Science to generating The Human and Ideational Sciences that define The Science of Freedom.**
>
> **"In this context, there is nothing more powerful than the human brain enriched by possibilities experience... All of his books and all of his demonstrations, collectively from the earliest to the latest, are the products of this processing phenomenon, profound alternatives for individuals, organizations, communities, cultures, and nations; indeed, for The Global Village and its Marketplace. The Sources of Freedom are Possibilities!"**
>
> (Cannon, Preface, *Science of Freedom,* 2007, p. xiii)

7. Generativity and the Human Sciences

Carkhuff's lifelong passion has been generativity or generative human processing. His recent work has focused on resolving the socioeconomic problems of our time through generative processing at all levels of community, culture, and economy:

- *The Generativity Solution, Building the New Economy.* Amherst, MA: HRD Press, 2009
- *The Generativity Solution, Volume III: Community Generativity.* Amherst, MA: HRD Press, 2009
- *The Generativity Solution, Volume IV: Cultural Generativity.* Amherst, MA: HRD Press, 2009
- *The Generativity Solution, Volume V: Economic Generativity.* Amherst, MA: HRD Press, 2009

Working with his colleagues, his body of work has culminated in a series introducing the **Human Sciences:**

- *Human Generativity: An Introduction to the Human Sciences.* Amherst, MA: HRD Press, 2013
- *The Human Sciences, Volume I: Probabilities, Possibilities, and Generativity Sciences.* Amherst, MA: HRD Press, 2013
- *The Human Sciences, Volume II: Probabilities, Possibilities, and Generativity Technologies.* Amherst, MA: HRD Press, 2013
- *The Human Sciences, Volume III: Carkhuff and Human Generativity.* Amherst, MA: HRD Press, 2013 (by B. G. Berenson)

Barry Cohen, Executive Vice President, Parametric Technology Corporation, summarizes Carkhuff's contributions as follows:

> **"His body of research has constituted the foundation for revolutions in all areas of human endeavor: human, information, and organizational resource development: government, corporate and community development; cultural, national, and now global economic growth. In short, he has changed the world by making social science a 'true science.'"**
>
> (Cohen, Foreword, *Generativity Solution,* 2009, p. ix)

Hernan Oyarzabal, Executive Director Emeritus, International Monetary Fund, summarizes Carkhuff's theory of the prepotency of generativity as follows:

> **"It is indeed the 'Generativity of Human Brainpower' and not the 'Economic Theory of Stasis' that hold the interdependent, enlightened, and entrepreneurial keys to our Prosperous, Participatory, and Peaceful future."**
>
> (Oyarzabal, An Open Letter on the Economy, *Generativity Solution,* 2009, p. xiii)

8. Summary and Transition

In summary, Carkhuff's body of work has differentiated him from all others in the history of science. Primary among his contributions has been the operationalization and application of **possibilities science and generativity.**

Carkhuff's greatest contributions may lie ahead—the transfers of possibilities science to his current list of generativity projects in a troubled world. It is left to his lifelong colleague, Bernard G. Berenson, an Einsteinian scholar, to place Carkhuff's work in historical perspective:

> **"Carkhuff's contributions to universal processing, alone, qualify him for leadership among the greatest scientists of history. His 'nesting, encoding, and rotating of processing systems' are the core processes of Nature's Generativity. In creating The Human Sciences, Carkhuff belongs in the Pantheon of Science along with the work of DaVinci, Newton, and Einstein."**
>
> (Berenson, B. G. **We Can Be Beams of Light.** Foreword, *Human Sciences, Volume II.* Amherst, MA: HRD Press, 2013)

In this context, Carkhuff is dedicated to the values of Albert Einstein, his exemplar:

> **"It is a very high goal: free and responsible development of the individual, so that he may place his powers freely and gladly in the service of all mankind."**
>
> (Einstein, *An Ideal of Service to Our Fellow Man,* 1950, p. 59)

Where Carkhuff joined the list of established contributors to Psychology in 1978, he now stands alone as the dominant force defining the **Human, Information, and Organizational Sciences** in the 21st century science and technology.

Along with his associates, he has architected and implemented new platforms for the following:

- **Generative Human Sciences**
- **Human-Centric Information Sciences**
- **Human-Centric Organizational Sciences**
- **Human-Centric Economic Sciences**
- **Generative Civilizations**

In so doing, Carkhuff and his associates have outperformed the entire body of the social sciences and provided leadership for the declining body of the physical sciences.

His colleague, Dr. Tom Kakovitch, a physical scientist, has the final word on generativity:

> **"Nature grows with generativity. That is why no one holds a monopoly over intelligence."**
>
> (Kakovitch, *Collegium,* 2012)

[1] Garfield, E. *The 100 Books Most Cited by Social Scientists.* Number 37, Institute for Scientific Information.

[2] Garfield, E. *The 100 Books Most Cited by Social Scientists.* Number 37, Institute for Scientific Information.

[3] Garfield, E. *The 100 Books Most Cited by Social Scientists.* Number 37, Institute for Scientific Information.

[4] Garfield, E. *The 100 Books Most Cited by Social Scientists.* Number 45, Institute for Scientific Information.

[5] Endler, Rushtore, and Rogdeger. *Productivity and Scholarly Impact.* American Psychologist, Vol. 33, No. 12, 1062–1082.

[6] Heesacker, Heppner, and Rogers. "Classics and Emerging Classics." *Journal of Counseling Psychology,* Vol. 29, No. 4, 400–406.

[7] Carkhuff, R. R. "Leader in Human Resource Development." *Education,* Vol. 106, No. 3.

[8] Berenson, B. G. and Cannon, J. R. *The Science of Freedom.* McLean, VA: American Noble Prize, 2007.

[9] Carkhuff, R. R. *The Age of Ideation.* McLean, VA: American Noble Prize, 2007.

carkhuff.com

mcleanproject.com

Printed in Dunstable, United Kingdom

91218907R00147